Wings of Courage

JACK D. STOVALL, Jr.

"WINGS OF COURAGE"

They were proud men,
    Flying through the bitter vale of combat;
Taking their chances,
    Against foe and splattering steel.

Not slowing for weather,
    They flew in raging storm and blinding overcast;
Offering their very lives,
    As the price of victory and freedom.

On they flew,
    Knowing not what tomorrow might bring;
Trusting in a higher power,
    They fought with a vengeance and yearned for peace.

They were brave men,
    Gallant in battle;
Magnanimous in victory,
    Flying on the wings of courage.

— By Jack D. Stovall, Jr.

In Memory of

Captain John Quinn West, Jr., USAAF
1919 - 1944
And to all the men of the 397th Bomb Group
who served their country with
courage and devotion.

"They gave of themselves that others might live."

Wings of Courage

JACK D. STOVALL, Jr.

PUBLISHED BY
GLOBAL PRESS
2990 Watson Road
Memphis, TN 38118

Copyright © 1991 by Jack D. Stovall, Jr.

All rights reserved. No part of this book may be reproduced or transmitted in any form or by any means, electronic or mechanical, including photocopying, recording, or by any information storage system, without permission in writing from the author.

ISBN 0-9615206-9-8
Library of Congress Number: 90-86361
Published by Global Press, 2990 Watson Road, Memphis, TN 38118

PRINTED IN THE UNITED STATES OF AMERICA

TABLE OF CONTENTS

CONTENTS 9

ACKNOWLEDGMENTS 10

FOREWORD 11

INTRODUCTION 13

CHAPTERS
1. RIVENHALL, ENGLAND, 1944 15
2. BACK HOME, 1941 28
3. MISSISSIPPI STATE UNIVERSITY, 1941 34
4. U.S. ARMY AIR FORCE PILOT TRAINING, 1942 41
5. MACDILL FIELD AND THE MARAUDERS 46
6. THE LONG FLIGHT OVERSEAS 65
7. COMBAT MISSIONS, LE PLOUY FERME TO OUISTREHAM .. 95
8. THREE DAY PASS TO LONDON 115
9. COMBAT MISSIONS, MANTES GASSICOURT TO ROUEN 122
10. FLAK LEAVE IN SCOTLAND 138
11. COMBAT MISSIONS, ST. LO TO NANTES 152
12. LONDON AND THE AMERICAN EAGLE 160
13. COMBAT MISSIONS, CLOYES TO CAUMONT 171
14. THE LAST MISSION 174
15. THE BAIL OUT AND ESCAPE 183
16. GOING HOME, 1945 191

THE FAMOUS MARAUDER PHOTOGRAPH 193

ABOUT THE AUTHOR 194

POSTSCRIPT 195

EPILOGUE 200

ACKNOWLEDGMENTS

A finished work is not nearly so much an achievement of the author as it is the work of those who have helped him, and I am grateful to the multitude of those who have given information and a helping hand.

To Bruce Stait of Cheltenham, England, who has always been most gracious to give research efforts and encouragement. Our correspondence has led to a deep and lasting friendship.

Then to the men of the 397th Bomb Group for their help with stories about Captain West and statistical information. Especially Neil McGinnis, Fred Daoust, Chester Natanek, Harold Zola, William Henry, Joe E. Jones, Nevin Price, and many others too numerous to mention. My deepest appreciation to each one.

Also to those who have allowed me to use their photographs and scrapbooks as reference material, I am greatly indebted. Particularly Raymond Snow and James M. Snow who so kindly entrusted me with their collections.

To Howser Hall, of the 322nd Bomb Group, a friend who has helped me with his vast knowledge of the B-26.

Then most importantly, to my family and to my wife, Patricia, who has been a source of unfailing encouragement through the many years of research and writing of this work.

To all of these and hundreds of others who have helped me, I give my sincerest thanks and deepest gratitude.

FOREWORD

For those of us who lived through the eventful years of the Second World War, life in wartime Britain can be as fresh in the mind as though it were only yesterday and all that is needed is to hear the sound of a siren or catch a glimpse of a photograph in a book, long forgotten. It may be fifty years ago but time-travel is surprisingly easy, given the right key. This book will no doubt enable many of its readers to make the backward leap to a time when great events were shaping the destiny of the whole world and Great Britain and the United States marched shoulder to shoulder against the forces of tyranny.

In 1944 I was a schoolboy, living in a small village in Essex, within easy cycling distance of three or four of the newly built airfields which, in the summer of 1944, were all occupied by units of the United States Army Air Forces - either the heavy bombers of the 8th A.F. or medium bombers of the 9th and my friends and I spent a great deal of time in finding ways of getting as close to the aircraft as possible. We were "mad" about aeroplanes. We just had to be near them and listen to the crews working on them and watch them take off and land. We seldom spoke of anything else that summer. Later on we began to take a keen interest in girls and some of my friends lost their love for the world of aviation, but somehow I never did - although ultimately I did find time to marry and raise a family. My wife may have had an inkling of what lay in store when I took her to the newly opened London Airport at Heathrow for a visit and proposed to her amid the roar of aero engines as Lockheed Constellations and Douglas DC-4's taxied close to the public enclosure.

The airfield at Rivenhall was less than a mile from my home, so it was only natural that I should spend far more time there than at the other bases. The gleaming, silver Marauders were a sight to behold and I never tired of watching them. Often before breakfast, we would see the squadrons take to the skies in the early morning sunshine, the thunder of their engines bringing many people outside to stare as the heavily loaded planes strained to gain height before setting out for the targets in Occupied France.

As incredible as it might seem, one of the aircraft I fancied the most was one named "By-Golly." My younger brother had ridden his bicycle several times to the aerodrome and had gone through a break in the hedgerow to talk to the crew of this particular aircraft.

Years later, I began to search out the men who had flown from Rivenhall and to record the history of the airfield for posterity and it was then that I encountered Jack Stovall. Can you imagine my surprise to find out that he also had a close tie with the plane named "By-Golly?" Our mutual interest in the units that were stationed at Rivenhall led to a life long friendship and I am honored to have the chance to write this Foreword to his book. The author writes of the heroism of the young men who fought in a war which lay many thousands of miles from their homeland in a tiny country which, without their help, would surely have been engulfed in a Nazi Europe. These young men were called upon almost every day to pit their skill and luck against an opponent they rarely saw, in a frightening world of flak and fighters in a bewildering kaleidoscope of bridges, marshalling yards and VI launch sites. They often had to limp back to base with damaged engines and with wounded crew aboard who urgently needed medical attention and the knowledge that in a few days — perhaps tomorrow — they would be asked to do it all over again.

These same young men enjoyed their all-too-brief excursions into the neighboring towns and the Essex countryside when they could relax from the horrors of the flak-filled sky and whistle at a pretty girl as she cycled by. Some of them even thought of a time when it would all be over and they could return to the U.S.A. and pick up the threads of their lives anew. They wrote home to their loved ones and tried to reassure them that their own part in this great drama was of small significance and that the risks were almost non-existent.

The men of the 397th Bomb Group were perhaps luckier than many of their brothers who had also flown in the dangerous skies over Europe, for many of them did return to their families who had patiently waited at home, hardly daring to pick up the mail lest it should contain the news they feared most. And eventually, at their reunions, they recounted their exploits to ' their comrades, who listened with many a sidelong glance and with an air of gentle disbelief. They have stories of parties in London, life in a P.O.W. camp, and a true tale of life with the French Resistance that would seem to have sprung from the fertile imagination of a Hollywood writer! But as the years pass the stories become a little more dim and the circle of comrades inevitably decreases at each reunion.

Those who did not return lie in foreign soil or perhaps are just a name on a wall of remembrance. But they are not forgotten. They are remembered each year when the nation pays homage to the fallen and their names live on in the hearts of their families and the men who fought with them so long ago. The sacrifice which was made by those brave men can never be forgotten. For without it, our world would be a far, far different place.

                    Bruce A. Stait
                    Cheltenham, Glos. GL 530BH, England
                    April 18, 1989

INTRODUCTION

This is the story of an American airman in England during World War II. It is a story of his hopes, heroism, and concern for his crew in the thick of combat. He piloted the medium bombers of the 397th Bombardment group on mission after mission over the "flak-laden" skies of German occupied France during the months before and after "D-Day" when the Allied Forces were fighting a determined battle for the second front.

The United States Army Air Force Bomber Command was trying with every effort to cut off the German supply lines by bombing bridges, railroads, and supply depots. Thus, the ultimate success of their bombing missions was won by the valiant young men who were willing to put their lives on the line for the sake of their country.

Their thoughts were of home, of families and loved ones, but sadly, so many were called upon to make the supreme sacrifice and never return to the homeland they loved so much. Their courage and sacrifice were typical of the Units of other freedom-loving nations who fighting beside them paid the price for the freedom we now enjoy.

Lest we forget ... The name of their Unit was "Bridge Busters" ... The name of the men was "Courage."

From left, Lt. Hughes, Capt. West, and Lt. Hunsicker. Kneeling is Sgt. Zola and sitting are Sgt. Robinson, Sgt. Picklesimer, and Sgt. Natanek.

CHAPTER ONE

RIVENHALL, ENGLAND, 1944

The men were beginning to gather in little groups as they waited for the briefing session with Colonel Coiner. There was always a sense of nervousness during these times of waiting as some tried to joke and make small talk while others just listened quietly turning their thoughts inward. They preferred getting on with the work of flying rather than wondering about target areas or how heavy the flak would be over the enemy territory. But it was a necessary evil; the waiting before flight time.

The lead crews had already been through an earlier briefing to finalize the main structure of the mission. There were notations of navigational headings, fighter rendezvous points, emergency radio frequencies along with altitude and time settings over checkpoints. Everything had to be exact. Every minute detail had to be checked and integrated in the overall plan.

This was the second mission of the day. There had been an early morning bombing of a railroad bridge at Boisset La Londe, and many of the crews who had flown the morning bomb run were rescheduled to fly this later one.

There was some dissension about double schedules becoming routine. For weeks before "D-Day" they had flown some doubles in preparation for the vast landings of Allied troops on the Normandy beaches. Now, even weeks after the troop landings, the Marauders of the 397th were still doubling some missions and pounding bridges, railroads and fuel dumps anywhere in France where they could immobilize the German enemy.

One of the officers of the group spotted Col. Coiner and his executive officer leaving the mess hall so he quickly alerted the others.

"Okay, men, here they come. Everyone into the briefing room."

The men filed in and took their seats just minutes before Col. Coiner strode into the building. When he appeared, one of the officers in the back called the group to attention.

Coiner moved down the center aisle and onto a platform in front of the men.

"At ease, men. I'm going to be brief. This is a combined bombing operation from Ninth Corps Headquarters. The 323rd, 387th and 394th are joining with us on this one. Our target is a railroad bridge at Nantes. Take a look at the coordinates here on our map. Your lead navigators will get you on target and signal for the drop. Captain Wood, you will lead the first box of eighteen aircraft. Captain Garretson, you take the second box. Captain West, you take back-up lead position behind Wood and take command of the first box if necessary."

Captain West's crew gave each other a familiar glance. They were proud of their Captain and proud of their valued performance on previous bomb runs. Their ship, the "By-Golly," had been selected as lead or back-up lead on nearly every mission they had flown.

It was no easy job to fly in the lead position which required a greater degree of skill. There were extra crew members and electronic gear to carry. It meant greater responsibility for the navigator, pilot and bombardier as the other planes would drop their bomb loads on the lead bombardier's signal. The extra weight was always a problem for the B-26. She flew nearly as fast as a fighter with high speeds ranging to 300 miles per hour at lower altitudes, but it took skill to get that heavy load off the runway and get her back home again.

Col. Coiner continued his comments at the briefing.

"Take off signal will be a green flare at 1842 hours, zero hour is 1900. Fighter rendezvous at St. Catherine's Point; zero hour plus 10 minutes. Altitude, 8,000 feet. Your coordinates are Base to St. Catherine's Point to target and retrace route to return. You can expect some heavy flak over the target area. Nantes is encircled with German 88's. Weather is favorable over target with 1/10 cumulus at 3,000 feet, haze up to 8,000 and visibility of 2-5 miles. Men, this railroad bridge at Nantes is a vital supply link to Paris from the coastal area, so let's get those bombs on target. Any questions?"

The men understood well enough and looked at him expectantly.

"You'll be carrying four 1,000 pound general purpose bombs with a 1/10th second fuse delay in the nose. This is a steel reinforced, concrete bridge so we will need some penetration before explosion. Good luck."

The men shuffled to attention as Col. Coiner left the podium. Meanwhile, there was the usual hubbub of conversation as the men discussed some of the difficulties of the newly assigned mission. Nantes was a tough place to be flying over and most of the men realized that fact. The 397th had seventy missions to their credit and they had long ago become familiar with which areas of France were heavily fortified and which were not. There was no such thing as an easy mission, but some were less difficult than others; the prime concern for the men was flak concentration over the target area. They had a way of getting around coastal flak sometimes because they developed a method of alternating their course every few minutes to evade flak before reaching the target. But, when the actual bomb run started, it was a straight line course and the pilot had to fly it hoping the flak wouldn't hit a vital part of the aircraft.

Many planes had flak damage on nearly every mission they flew and still made it back to the base. It was amazing to see the flak damage a Marauder could survive and still continue to fly. Severed hydraulic lines, cut control cables, gaping holes in wings and fuselages, rips in gas tanks and hits in the engines could be sustained sometimes without knocking the ship down. Yet, other times a hit in the wrong area was deadly, particularly in the engine compartment with an ensuing fire. Sometimes a crew could limp home on one engine, other times it was worse and the pilot pushed the bail out alarm hoping the men could all get out before the Marauder went into a violent roll, spin or dive. Invariably, this pinned them in the plane leaving no way of escape.

Captain West walked out after the briefing with his crew.

"Lt. Budge, did Robbie say the ship checked out okay?"

"Yes, Sir. She's all right."

"Okay men, get your flight gear and meet us at the squadron area. It's thirty minutes before take off," he commented as he glanced at his watch.

It didn't take long for the men to get their chutes and flight gear. When the trucks pulled up they were throwing in equipment and piling in to go down to the Squadron perimeter.

The ground crews had already been working for hours to prepare their aircraft for the mission. Engines were checked; hydraulics and control systems checked. Armorers had loaded the fifty caliber ammunition for the guns, and the four 1,000 pound bombs had been pulled into the forward bomb bay, their fuses set and the electrical bomb releases connected. It was a tedious operation as the crew chiefs fussed over their aircraft like mother hens.

It was a matter of pride for each crew chief to think they had the best ship in the squadron, and when the flight crews arrived it was "thumbs-up" and "she's ready to roll." The flight crew's truck had pulled in and unloaded the "By-Golly" crew with their chute packs. Captain West was checking his men carefully.

"Zola, did all the crew make it over on the truck?"

"Yes, Sir. Nat and Picklesimer are in already, and Webb and a new man are still unloading their gear."

"Who is the extra man?"

"Sergeant Joe Jones from the replacement pool. He'll be flying as waist gunner this trip."

"Lt. Daoust, is Cramer in?"

"Yes, Sir. He and Budge have already gone to their stations."

"Okay, let's get in and get rolling."

Lt. Fred Daoust was the crew's bombardier and one of the best. He could find targets that others had trouble spotting, but that was his job. He was a young man and quiet, but an expert with the Norden bombsight.

West followed Daoust as they climbed through the nose wheel well and into the cockpit area. Lt. William Budge was already in the co-pilot's position checking instruments when West came in and took his place in the first pilot's seat on the left.

"Bill, we'll follow Captain Steere on roll out. Is everything checked out okay?"

"Yes, Sir. The ground crew have hooked on the external power and we're clear to start number one engine."

Captain Quinn West slid the side window open and took a glance at the port engine.

"All clear, Robbie?"

"Yes, Sir. She's clear."

The inertia starter began turning as it joined the rumbling, explosive noise of other squadron aircraft starting their engines. The big twelve foot, four-bladed prop began to turn slowly at first like a lazy fan, then faster while clumps of smoke billowed out the exhaust stacks as she caught and began to turn on her own power.

It was a beautiful sound to hear, that 2,000 horse power engine level out and ease into a steady 1,000 RPM. There was just something about it that always gave a thrill with the sense of power when pushing the throttles forward and feeling the response of those great engines.

Now the engine noises were like thunder as all the squadron's aircraft began to add their power, and one by one the planes taxied down the perimeter to the main runway. West audibly called the instrument settings as he glanced at the panel before him.

"Engine temp and manifold pressure okay, cowl flaps open, carb. heat and gas gauges okay. Bill, set the flaps at 20 degrees."

Quinn angled the plane toward the main runway as he braked the forward motion and began to run the engines to 2,400 RPM. The prop wash from his engines and the engines of those ahead of him were shaking the plane like the oncoming winds of a storm. The engines slowed as he pulled back on the throttles, then he glanced toward the tower waiting for the green flare. Captain West's voice came over the intercom.

"Men, we're about two minutes from take off. Let's breathe a silent prayer for a safe mission." There were a few moments of silence. The men were always glad for the opportunity of silent prayer. It relieved the tension of waiting for take off and it helped to calm their fears knowing that a higher power was guiding their lives.

"There's the signal flare. We're cleared for take off."

The lead planes began to move onto the runway and one by one at 30 second intervals, they pushed full throttle and began the rush to get airborne. Quinn saw Captain Steere moving out and he followed slowly. Then after lining up on the runway, he steadily pushed the pair of throttles forward. The acceleration was beginning to press him back into his seat as the air speed indicator began to move and show his airspeed. 100, 120 miles per hour and the runway markers were beginning to blur in his side vision. At 120 miles per hour, West eased back on the control column and the nose wheel broke ground. 135 and she was getting light on the main gears, but nearly a mile of runway had been covered. He hauled back on the controls and the plane began to lift.

"Hit the gears, Bill."

The aircraft hung on the props across the boundary markers and just cleared the tree line at the end of the runway.

Zola whistled, "Whew! She's loaded down today, Captain. That was a close one but she's airborne."

The forming up process was routine as the leaders circled and waited for the aircraft following to form into flights of six navigating into the box conformation, then move onto their planned heading. There were two boxes of eighteen aircraft each on this mission, and as usual they were flying at slightly different altitudes. The "A" box was high, and the "B" box was low. West's aircraft slipped into the slot behind Captain Wood and backed him as Box "A" deputy leader. The flight was ten minutes out and nearing St. Catherine's Point when the fighter escort was sighted.

Zola called over the intercom, "There they are, Captain, at three o'clock high. Looks like 9th Air Force Mustangs."

"Okay, men, you can clear your guns when we get a little farther over the Channel. Radio silence from here on in."

It would be a waiting game now, traveling across the Channel. It all looked so peaceful at 8,000 feet with a few ships scattered here and there. One could almost believe all was at peace, except for the tight gut feeling the men had, knowing that soon the coastal, anti-aircraft batteries would be firing at them and trying their best to knock the Marauders out of the sky.

Sergeant Zola could see nearly the whole flight of 36 aircraft stretched out behind their plane from his position as tail gunner. It was a beautiful sight, but he didn't have long to think of the skillful symmetry of the group for he was a crucially busy man. His job as engineer gunner was an important one, being in charge of the aircraft's function and maintenance while in flight. He had to be well trained in the fields of engine mechanics, hydraulics and structural engineering. When anything went wrong with the aircraft it was, "Hey, Zola, see if you can find the trouble and fix it." The "By-Golly" was his baby and he knew how to keep her in tune.

Lt. Daoust's voice came over the intercom, "Captain, start evasive action. We're nearing the coastal batteries."

They employed a zig-zag course changing direction every twenty seconds to keep the German gunners from zeroing in on the Marauder formation. The process would continue to the "Initial Point" or "IP," then the bombardiers required a straight line of flight to the target establishing position by picking out points of reference on the ground that would keep them on course.

Daoust's voice sounded again, "Captain, we've reached the IP, keep her straight and level ... bomb bay doors open ... we're on PDI."

No sooner than Daoust had spoken, flak began to burst everywhere. It was heavy and accurate for the flak bursts were buffeting the ship in every direction. Black puffs of smoke hung all around from near misses. The close bursts often rattled shrapnel over the ship, some penetrating the aircraft, others exploding with enough concussion to knock the ship off course. Captain West watched the PDI instrument on the panel to keep his ship on course during the bomb run. The PDI was linked to the bombsight and indicated the aircraft's drift from the line of target.

The flak continued relentlessly and minutes seemed like hours as they wondered if a flak burst might disable the ship or kill some of the crew. Fred Daoust was bringing the ship on target even through all of the dangerous confusion. His eyes were fixed on the Norden view finder as he concentrated his full attention on picking up the target. Finally, Daoust's voice broke the silence of the intercom ...

"BOMBS AWAY!"

The ship gave a sudden lift upward and no one had to ask questions. They knew the bomb load had been dropped. The "By-Golly" banked suddenly to the left and dropped altitude picking up speed in the turn. West was putting the ship on her return course.

"Daoust, what does the bomb drop look like back there?"

"Looks like we are right on target, Capt'n. There goes the main blanket, and the bridge is covered with smoke. Hey, take a look at that! She's broken in the middle and dropped a span in the river."

"Fred, you must have put the whole first box on target. Man, what a sight.

The whole bridge has swayed over to the right."

They had just begun evasive tactics when suddenly the flak became heavy again. The black puffs of exploding shrapnel were right on them. West was turning to evade the flak's direction, but it wasn't working, the bursts followed each turn they took.

"Bill, this flak is terrible. They must have plotted our range perfectly. These bursts are right on us." The ship took a sharp jolt to the left.

"We've taken a hit somewhere. Bill, check the instruments."

"There it is Capt'n, smoke pouring out of the starboard engine."

"I'm pushing full throttle on number one, trim in all the left rudder we've got. It's still yawing to the right. Hit that extinguisher on the engine. Maybe we can control the fire."

"I don't think it's working, Capt'n. That nacelle is blown up too bad."

"Trim in some left aileron, she's still pulling to the right. I'm cutting switches and feathering the right prop. We must be hit somewhere else 'cause she's losing too much altitude. Zola, what's happening back there?"

"Captain, there's some hydraulic lines cut, fluid is all over the place. Plenty of shrapnel holes, but no major damage that I can see now."

"Are any of the men hurt?"

"No, Sir. They are all okay. Hey! There's gas pouring out of our right wing tank. Can Bill see it from up there?"

"Yeah, he says it's gushing out. If the gash is too big, the self-sealant won't work. Zola, the instrument panel has gone out on us so I can't tell anything about the fuel gauges. If it don't get better fast, we're going to have to bail out."

"Capt'n, the gas has stopped flooding out of the wing tank, but I'm afraid that right engine burning might catch it all on fire."

"Bill, she's flying so mushy it feels like she's nearly at stall attitude. Zola, get some weight out of that tail section. Get Pic and Lloyd to help you throw everything out that's not nailed down. Throw out the guns, ammo, flak suits and everything else you can find except your chutes."

"Nat, break the radios apart and throw them out. Anything else you can find, throw it out, too. Bill, I don't think we can make it across the Channel on this one engine, and if that gas keeps pouring out, we had just better call it 'quits' and bail out. If we could just find a field, I think we might be able to 'belly-in.'"

"I don't know, Capt'n, it sure would be tricky trying to go in with that leaking gas. This whole thing might blow up."

"Zola, give me a report on what's happening back there."

"Sir, the gas has stopped pouring out of the wing tank, and we've thrown everything out we possibly could. It's still a mess back here with hydraulic fluid everywhere. Capt'n, I just saw a couple of Mustangs escorting us down."

"Good, maybe they can keep the German fighters away 'til we get out of enemy territory. Doug, do we have any airfields close for an emergency landing?"

"Yes, Sir. My maps show a Field near the Normandy coast that's operational. It's marked 'A-7' and probably a fighter base."

"How far away is it?"

"About five minutes on our present heading. It should be off to the left if we hold our course."

"Bill, I'd give anything to know how much gas we have in those wing tanks. Without the fuel gauges it's a risky gamble trying to belly-in worrying about losing the engine right at the most crucial time."

"Yeah, I know, but she is riding a little better since we dropped the extra weight. Look! ... There's the Airfield ... off to the left. That runway sure looks good."

"You bet it does ... All right men, we're going to come in for a wheels up emergency landing. You all know the procedures. Come up to the forward bulkhead and get set for a rough one ... there she is, Bill. Think we can make it?"

"I don't know, Capt'n. I just hope we have enough gas for a couple of approaches."

"Okay, let's try it."

West began to pull back on the throttles and start a slow long turn into the final approach. During those crucial minutes he began to call instructions to his co-pilot and talk to himself as well as he carefully articulated each step of the emergency procedures.

"She's lining up pretty good ... Crank in full flaps ... Easy old gal ... Don't stall now ... Just ease on down ... Easy, now ... That's the way ... Hey! What's that truck doing on the runway down there?"

"I don't know, but look! The tower is waving us off."

"They can't do that! ... We'll never make it around again. Cramer, shoot some more emergency flares ... Don't those guys know we're in trouble? ... Bill, trim in all the left you can ... I'm pushing full power ... Crank a little 'up' on those flaps, Zola ... Left rudder ... Easy, she's lifting ... She's coming up a little ... She's shuddering, but she's pulling up ... Good, now just a shallow turn slow and easy ... Hold on girl ... Don't stall on us ... We've got to keep easing her to the right, all the way around again to line up ... Okay, I'm pulling back just a little on power ... Get ready to cut the switches before we hit ... One spark could blow us to 'kingdom come' ... Zola, crank in some more flaps ... Easy now, she's getting a little mushy ... Hold her steady ... Hold her."

"Captain, the tower is waving us off again. I can't believe it!"

"No way, Bill. We're going in anyway ... We can belly into the runway and slide off the side when we slow ... Here we go, fold up into a knot, men. We may hit hard ... Easy, let her settle ... Easy now ... Cut the switches! She's skidding ... Hard left! Hard left! ... She's holding together, Bill ... Hold that left wing up ... She's skidding straight ... She's holding ... We got it made! We've got it made now! ... CLEAR OUT, MEN!! ... Clear out fast! ... She's smoking inside ... Zola, get 'em out! ... The right wing's on fire ... Make sure they all get clear of the ship ... Bill, open the top of the escape hatch and let's get out of here! ... Where's Fred and Doug?"

"Here they come, Capt'n."

"Let's clear it men ... Come on out the top hatch ... Jump and we'll follow you out!"

Dense, grey smoke was beginning to pour out of the cockpit and engine compartments as the men jumped out and moved back fearing an explosion. Flames were now engulfing the right engine and wing area with heat building to an intensity that made the men turn away, then stand back further away from the burning aircraft. The entire crew watched sadly as the whole front section of the aircraft began to crumble. The flames were spreading from the wing areas and fuselage midsection as the emergency crew with equipment began to try and contain the flames with foam.

"She's really starting to burn now. Look at that right wing, Capt'n. Let's get away from this thing! ... She might blow any second."

"Zola! I'm going back in there and see if all the men are out!"

"No you aren't, Sir. She's burning too bad ... We're all out ... I swear it! I've counted 'em ... Everyone is out."

"I just don't want to lose anyone!" Quinn yelled as he looked around frantically. " Sgt. Jones?"

"Here I am, Captain," West heard from behind him. " We all made it out okay."

More crash trucks and ambulances were beginning to arrive as the men moved back to let them through. Flames were now out of control and the men slowly gave up the "By-Golly" to fate.

"Okay, men, stand back and let them come in ... Give 'em room to work ... Nat, I guess what the fire don't ruin, the foam will ..."

"Captain ... I hate to see her burn like this ... I was hoping we could go through the whole war with her, and all our gear is in there, too."

"I know, but we're all safe and that's really what counts. God was good to us, Nat. He brought us through a real tough place today," Quinn commented sadly, but thoughtfully.

"Yes, Sir. I felt like we would make it, but when she started shivering after that first approach, I thought we were finished."

"Yeah, I was having a few doubts just about then, too ... Stand back, men ... Let them spray the foam on ... Maybe they can salvage some of our stuff inside."

"No, Skipper. She's gone ... 'By-Golly' is gone ... She's burned too badly ... I don't think we will ever be able to salvage anything out of her now."

While they were talking more trucks were beginning to arrive filled with men from the Air Base who were curious about the bomber crash.

"Zola, it looks like some of those men in the trucks know us. They're waving and shouting like they do ... Do you know any of them? ... I can't believe it! It's your brother? How in the world did they know it was your plane?"

By this time, Zola's brother had jumped off the truck and ran out to them as he exclaimed, "I saw that big yellow stripe on the tail, Captain, and I just knew it was him, Sir! I just knew it!"

Zola and his brother were elated and continued to slap each other on the back while asking about family and friends. The reunion of brothers seemed to add a needed happy note to the untimely plane crash as the "By-Golly" crew loaded on the trucks to ride to the other side of the Air Base.

Zola grinned, "Captain, what do you think of my brother, Dox, being out there to meet us when we came in?"

West reflected for a few seconds and smiled, "By golly, that's really amazing!"

Zola's brother began to tell them that the Commanding Officer was in a rage because they had torn up some of the steel matting during the crash landing.

West said, "You must be kidding! How could anyone be mad because of an emergency that could have killed us all?"

"I don't know, Captain, but you know how those fighter pilots are about their runways... Did they tell you that the truck on the runway was defusing a bomb?"

"No, Gosh! No wonder they were waving us off." The trucks pulled up to the Operations Building and the men began to pile out. West went inside warily to report to the Operations Officer.

"Sergeant, is the Operations Officer in?"

"Sir, I think the Commanding Officer wants to see you first. His office is on the left."

Captain West opened the office door and saluted the Colonel, who was looking at reports on his desk. He turned and eyed Captain West critically. Without a greeting he began to address him.

"Captain, wasn't there any way you could have made it back to your base on one engine?"

"No, sir. We had flak holes in both wing tanks and our panel wasn't working, so we had no way of knowing how much gas we had left. I wasn't sure we had enough to make it to this base."

"Well, why didn't you just bail out? I can't believe you tore up our runway so badly. It will take us a month to get more requisitions filled for those steel mats."

"I'm sorry, Colonel, but bail out is the last resort for bomber crews. We were over enemy territory until we sighted your base, and I don't like to think of my men being put in German prison camps. It's different when you fly bombers because your crew depends on you, and it's important not to separate them by a bail out. I know most of my crew would rather risk an emergency landing together than bail out and be separated."

"Okay, West. If you say so. Anyway, congratulations on the crash landing you made. I don't think I could have pulled it off in that 'hot shot' bomber you guys fly."

"She's not that bad, Sir. If we get back in a good ship some day, I'll take you up."

"No thanks, West. I wouldn't risk my life in that ship. I'll just stick to the Thunderbolts. They're dangerous enough."

"Colonel, when can we get transportation back to our base?"

"Well, our radioman has contacted Rivenhall, and they will probably send a ship to pick you up. We don't have any C-47 Transports assigned to us."

"How about quarters, Sir?"

"That's a more difficult problem. If you don't mind sleeping in pyramidal tents, I think we can find something for you."

"Thank you, Sir."

"West, sorry I blew my stack about the runway, but that deck is precious to us."

"It's all right, Sir."

Captain West walked out of the Operations Building and began looking for Cramer and Budge. He found them talking to some of the crew outside the mess hall so he walked over to the group.

One of the crewmen asked, "Captain, where have you been for the last hour?"

"Getting blistered for tearing a runway apart. Have you guys eaten chow yet?"

"No, Sir. We've been waiting around for you."

"Well, let's see if we can't get something to eat. I'm hungry as a bear!"

"We are too, Captain, especially after that last landing."

"Wait a minute, I thought you guys liked my flying!" Quinn laughed. "What about it, Zola?"

"Well, I was beginning to have my doubts there for awhile." Zola was grinning and all the others started laughing.

After chow, the men began to talk about the possibilities of getting a bunk for the night. The danger of the emergency crash landing had left them exhausted.

Daoust spoke first. "They don't have an aircraft that can get us back to Rivenhall; so what do we do now?"

Nat answered, "Captain, Sgt. Jones and Pic have found a barn near the base that they are going to sleep in."

"That sounds fine to me, but Cramer, Budge and myself are going to bunk in with some of the officers in the pyramidal tents. They're not great accommodations, but they are close to the mess hall. Don't get into any trouble, Nat, and you guys meet us here in the morning for chow."

The "By-Golly" crewmen turned and ambled in the direction of the old barn as Quinn and his officers walked to the tents. It had been a long day, and they were all tired. The initial crush of events had exhausted them as one thing had piled upon another to complicate the situation. Everything seemed to be happening so fast it was unbelievable.

Quinn found a bunk in one of the tents and laid down to close his eyes for a few minutes, not even bothering to take off his flight jacket or boots. It was too much of an effort. Just then Lt. Budge came in to stow his blankets on the other bunk. West asked him drowsily about the men.

"Bill, are all the men bunked in?"

"Yes, Sir. They're fine."

"Good. Be sure to check with Operations so they will try and get us transportation back to England."

"Yes, Sir." Bill quietly went out of the tent and closed the tent flap. He knew that West was "dog tired" and it amazed him that West's last thoughts were always about his men.

After Bill left, Quinn allowed his thoughts to drift back home as he sometimes did before dropping off to sleep. It was at these times that he could block out all the tension and battle weariness of each day.

He closed his eyes and could almost smell the fresh scent of new mown hay and feel the healing warmth of an early morning sun slowly moving over his tired limbs. Always with sleep came those dreams of home. How real they seemed. It was a precious escape from the waking reality of battle.

From left, Lt. Budge, Lt. Cramer, Capt. West, Sgt. Jones, Sgt. Webb, Sgt. Zola, Sgt. Natanek, Lt. Daoust, and Sgt. Picklesimer.

CHAPTER TWO

BACK HOME, 1941

The early morning dew was heavy and clung to the meadow grass like droplets of shimmering crystal. One could see every track of the bird dogs up ahead and many tracks of smaller animals which had criss-crossed the open fields during the hours before dawn.

Two hunters were walking slowly just over the top of a low hill. The older boy was ahead and carried the gun. The younger boy was trailing behind several yards, jumping from track to track which had been made by the footsteps of his companion in front. The hunter looked back with a grin.

"Hey, big boy, hurry up. I think the dogs have found a covey of quail."

He noticed the kid trying to jump the long steps from one track to another and he called back to him again.

"What are you doing, Jackie?" Quinn laughed.

Jackie had a look of deep concentration as he replied, "I'm walking in your footsteps."

"Well," Quinn said, " you're goin' to have to grow some more to step that far. Come on, let's try and get those birds."

The boys slowed their pace now, walking quietly over to the dogs that had caught scent of the quail ahead. The dogs moved in slow motion like they were frozen, slowly picking up a foot and waiting, going another step farther, then holding still on point with bodies stiffened and senses aware the birds were close.

Suddenly, there was a startling noise of beating wings and a flurry of movement breaking the silence of the fields. Quickly a deafening shot rang out, then another, and two birds fell from the flying covey. At the hunter's signal, the dogs bounded in and retrieved the birds to carry them back for a rewarding pet. It was a game that never tired the dogs.

Jackie's face reflected the excitement of watching Quinn's hunting ability, and he exclaimed loudly, " Gosh, it scared me when those birds flew out. How do they make so much noise? Didn't it scare you?"

"You get used to it after you've hunted for awhile."

"Do you think Aunt Jo will cook them for supper tonight?"

Quinn hesitated, " Well, if we get on home and get them dressed, she might."

The pair walked back toward the house together while Jackie's voice became quizzical and pleading.

"Quinn, will you show me how to shoot a gun someday?"

"Sure, you can probably handle this 410 gauge."

That seemed to satisfy the small boy immensely, and he settled down to the task of the long walk back home.

The fields were still now except the playing of the dogs far ahead as they sniffed every trail left by an earlier passing rabbit, and barked to encourage the other to follow. The morning air was cool and fresh with the smell of newly cut hay as the boys crossed the back pasture and angled toward the farm complex.

The rolling hills of the countryside were always a place of beauty for the younger boy who came there for several weeks in the late summer to visit his aunt and uncle. It was an exhilarating change from the confines of the city which meant the pressures of the school work he had to do. He knew that was necessary, but here it was like another world, with small town friendliness and warm people who really cared about each other. They had the luxury of taking time to enjoy life at its best. It was always a special treat whenever Quinn was home from his studies at college. He was more like a big brother to Jackie than a cousin because he took the time to take him on short trips into town and on hunting ventures like the one today. Quinn was cut from a different mold than others. Maybe it was his training in the Boy Scouts or his work with the young people in his church, but there was a genuine element of friendliness and caring that made everyone like him immediately. It was a natural trait of his personality to have a deep interest and concern for others.

Quinn and his young cousin came up a low hill from the pasture below, and the familiar outlines of the farm buildings began to appear. There were large silos with the storage barns beside them, a low roofed milking barn and the smaller cream separator building, then the long, low concrete block building used for the farm's blacksmith shop, and finally, the buff-colored, brick home; large, but unpretentious, came into view.

The buildings formed an outstretched semicircle with a large open work area in front, and as the boys neared this area they saw several workers attaching a mowing machine to the back of a large tractor. The men looked up when they heard the boys approaching.

" Mornin', Mr. Quinn. Law' me, you done took that boy huntin' with you. Did you get some birds?"

"Yep, we ran into a big covey down in the back pasture."

"Did Mista Jack get to shoot any?"

Quinn laughed, " No, we haven't turned him loose with a shotgun yet." The kid grinned, and broke into the conversation.

"But he's already promised to teach me how to shoot a gun."

"Well, I declare, that boy is into everything around this place," Tom said laughingly." We're gonna have to put a harness on him to keep him out of trouble."

Tom's good-natured humor was infectious, and everybody loved him. He was the head worker on the place, and there was never any doubt about his faithfulness to the people he worked for. He loved them like they were his own family. He was always happy with his work and hardly ever grumbled. It was just something that he undertook as part of the joy of living, this working on a farm. One could always tell he was near because even over the noise of a tractor engine, one could pick up the gentle spirituals that he would sing, hum, or whistle while he was working in the fields. It was a harmonious combination; not just words in clear phrases, but part humming with words intermixed. It was beautiful to hear the rhythmic monotone of the tractor's chugging and hear Tom's voice break out in song. "Yes, my Lord ... Yes, my Jesus knows." The words were simple, but to those around him, it sounded more beautiful than a " Hallelujah Chorus." He was just a humble man with a heart so full of praise for his Lord, that it had to overflow into song.

The boys walked on toward the house, giving a whistle for the bird dogs to come along. The dog pens were in back of the lot behind the house and were well built wire enclosures with a dog house inside to protect the dogs from the weather. Quinn opened the pen door and encouraged the dogs to run in.

"Come on, boy. That's a good boy."

As they turned to go to the house, Jackie began to ask in halting words, "Could we ... I mean ... Could you ... Show me how to shoot a gun this morning?"

"You don't mind getting a sore shoulder, do you?"

"Nope, I don't mind."

"Well, get that piece of cardboard and set it down for a target out away from the pens. Now, look at how I am holding the gun. Hold it firm and keep it tight against your shoulder. Stand like this, and lean forward as you sight down the top of the barrel ... Get the front bead on the target, and slowly squeeze the trigger as you hold your breath. Have you got it? ... Okay, let me load it for you and it's your turn. Be careful, and relax a little, now bend your knees some more ... That's it ... Keep your sights on the target, and slowly put pressure on the trigger ... Keep squeezing..." The quick explosion was deafening as the recoil was enough to make Jackie wonder what had hit him.

"Okay, let me unload the gun, and we'll take a look at the target," said Quinn protectively.

Jackie still had a stunned look on his face as he exclaimed, "Wow! That thing really kicks doesn't it?"

"Yeah, I think you shot a little low. There's just a few shot in the bottom of the target."

Jackie began to protest, " I couldn't help it. The gun was moving up and down."

Quinn laughed, " It takes practice, big boy; lots of practice. Let's put the hunting stuff away. I've got to get down and help Tom with the mowing today."

Tom was finishing with the mowing machine attachment as the boys came strolling into the work area.

"Tom, I'll put some blades on the Ford tractor and help you down in the back lot if you want me to."

"Yas, Suh. We can get thru' today with you helpin'."

The John Deere tractor looked like a giant to the kid, and he was spellbound by all the preparations for the mowing work. The tractor had a large flywheel exposed on the left side which was used for hand starting the engine. Tom was beginning to pull it through to start it. The motor caught for a second as clumps of white smoke coughed from the stack, and she began to hit solid giving the sounds of power. A person could tell the sound of a John Deere a mile down the road. It had a deep throated, even tempo which was different from other engines.

The motor was running evenly now, as Tom called over the engine noise, "Mista Quinn, I'll git started now, and you come on later." The tractor lurched forward as the clutch engaged, and Jackie watched wide-eyed, wishing every minute he could be on the tractor with Tom.

Jackie was too young to be allowed to help with the more dangerous work. Most of the time he was just in the way satisfying an untiring curiosity about the farm machinery and excercising a knack for getting into trouble. Last summer he had gotten too close to a powered conveyer belt which caught his sleeve in the gears, but Tom had been quick enough to jerk his arm free of the machine with nothing more serious than a torn jacket. The best way to keep him out of trouble was to provide some work for him to do, but the little jobs never seemed to last long enough and soon he was back in the barns, shops, and dairy building looking for chores to do.

As the noon hour arrived, the family had gathered for lunch and Jackie took his place beside his cousin. Quinn's sisters, Majorie and Josephine, were not at home this week as they were attending ceremonies at Blue Mountain College. Soon, Quinn also, would be leaving home to attend a summer camp R.O.T.C. at Fort McClellan, Alabama, before starting his senior year of college at Mississippi State.

After prayer was said, the table was served and everyone began filling their plates while Quinn was pouring Jackie's glass of milk.

"I don't want very much milk today," Jackie complained.

Quinn laughed and said, " Come on boy, you've got to grow some, and this is just what you need."

Jackie gave an unconvincing glance as he said slowly, "All right."

There was no use arguing. Everything on the table was wholesome from the roast beef and vegetables down to the slices of whole wheat bread and butter placed in the center of the table.

The meal time conversation was light with some talk of the mowing going on in the field and work set out for the afternoon. Jackie was getting restless and wanted to help with some of the work, so he broke into the conversation with a youthful plea.

"Uncle Jake, can I help Quinn mow this afternoon?"

"Well, I don't think you could help there, but you could pick some peas in the field across the road."

It pleased Jackie to get a special job and he almost shouted.

"Oh Boy! I'll get a big sack and pick lots of them."

Quinn smiled at the kid's enthusiasm, knowing that the summer sun would soon make a short job out of the work project. He suspected Jackie would be playing around the barn area by early afternoon. Quinn liked to have Jackie there with him occasionally to break up the somewhat humdrum pattern of farm work for him. Then too, the kid reminded him of himself as a youngster and some of the trouble he used to get into around the farm.

Jackie had fallen silent and seemed pensive. "Is it fun to go to college?" he asked suddenly.

Quinn replied, "Well, sometimes it is, but mostly it's just going to class and studying, about like you do with your homework in school."

"Gosh, I thought it would be more fun than that. I saw a soldier's uniform in your closet yesterday. Do you wear it all the time at school?"

"Not always," Quinn replied with a grin at his parents. "Just on special days for inspections and drill. It's an R.O.T.C. uniform."

Jackie got quiet again as the meal continued, and Quinn's father spoke, "Will you be packing tomorrow to leave for summer camp?"

"Yes, Sir. The classes won't start until next month at school, but the senior military units are meeting early for the camp."

"Do you have a ride down there?"

"Yes, Sir. One of the boys will pick me up."

The meal was ending as Quinn and Jackie asked to be excused to leave the table. Before going outside, Quinn stopped by his room to put on his work boots. Jackie followed him feeling privileged for the chance to share Quinn's work routine. The room was large with a separate closet that was forever cluttered with hunting clothes, boots, shoes, the guns he hunted with, and boxes of shotgun and rifle shells pushed to the back of an overhead shelf. Needless to say, it was a favorite rambling place for Jackie. Nothing pleased the kid more than to put on an old oversized hunting coat and some of Quinn's large hunting boots just to tramp around the outside of the house, acting like a full-fledged hunter.

After Quinn had pulled on his work boots, he looked at Jackie. "Come on big boy, we've got some work to do." As always the kid followed behind, but today he walked slowly, thinking all the while that tomorrow would come quickly, and Quinn would be leaving for school. Then it would be time for him to be sent back home and a whole year would pass before he could get back here again for another summer vacation.

Little did they realize then, that a far distant place named Pearl Harbor would soon change the leisurely summer vacations for them along with millions of Americans everywhere who would have to fight for their country's freedom in the years ahead.

"Billie Willie V" and her crew; Sgt. Doolittle, Lt. Jernigan, Lt. Duncan, Lt. Billy North, Sgts. Gerlach, Downs, Watermolen, and Hoit. Pilot, Wm. North is standing on right.

CHAPTER THREE

MISSISSIPPI STATE UNIVERSITY, 1941

The campus at Mississippi State was beautiful in late summer as the trees and shrubs looked more resplendent. Soon the fall season would color the landscape with a patchwork quilt design of reds and yellows.

Quinn made his way briskly across the quadrangle and angled toward Lee Hall. The buildings were all familiar landmarks to him after three years of residence. There were "Old Main," "Hull Hall," and "Magruder Hall" dormitories, as well as the engineering, agriculture, and botany buildings. All brought back a flood of happy memories, but more of his free time had been spent with activities at the YMCA building than any other place on the campus. It warmed him to think of all the students who had passed through its halls influenced by guidance and fellowship the association offered.

It sounded almost too quiet without the usual clamor of students and classes, but registration for the fall semester did not start until next month. The school would turn into a "bee hive" of activity soon enough. Only the ROTC cadet officers were assembling today for their annual six week training program at summer camp. It was to be held at Fort McClellan, Alabama, this year and offered an intensive weapons course, which the sixty officers had been looking forward to attending.

After completing this summer's training and their senior year in college, they would be given commissions as Second Lieutenants in the U.S. Army Reserve Corps. The work was going to be hard, and they would be living under field conditions, but the exciting part would be firing all the Army's weapons from the light machine gun to the heavy field artillery pieces.

As Quinn approached with a wide smile and a ready handshake, he joined the group of seniors who were already talking and waiting for the meeting to start.

"Hi, West," several exclaimed. "Where have you been since we saw you last at the Mississippi State Baptist Student Union Convention?"

"Just working, except for a trip to Ridgecrest last month."

"We thought for sure that you would win the presidency of the B.S.U. council this year. What made you withdraw your name from the nomination?"

"Well, I really hated to do that, but I wouldn't have been able to do a good job with it this year because of our summer camp, and the YMCA will take a lot of my time also."

One of the group exclaimed, "Hey, that's right. Quinn's going to be a big man on the campus this year with the YMCA presidency," and they all laughed approvingly.

"I don't know about the big man stuff, but there will be some real challenges in the 'Y' work," he commented.

A whistle blast was heard coming from the ROTC assembly rooms in Lee Hall as one of the men came out and told them the meeting was starting.

Colonel Rose began his opening remarks, "Men, you all have some personal equipment with you, and there will be some military equipment we will load on the trucks when they arrive. The Army trucks should be here about 0900 hours. I am proud of this Unit, and I expect each one of you to conduct yourself as an officer and a gentleman during this training period. There will be hard work to be done, and it's going to be hot down there especially on the rifle, pistol, and weapons ranges, but try your very best to qualify on every weapon as these scores will become a permanent part of your Army records. You will have some free time, of course, so enjoy it, but work hard when the training hours are scheduled. Okay, men, let's get some of this equipment outside and get ready to load when the trucks arrive."

After loading on the trucks the trip was uneventful except for the singing, horseplay, catnaps, and chatter of the cadets along the route. As they neared the camp, most of them were beginning to notice the terrain more closely. The Main Post at Fort McClellan looked good with paved streets, permanent living quarters, Post Exchange, and theaters, but the cadets were going to the outskirts of the camp bivouac area that had less of the comforts of home. Their area was a large, open, sandy stretch of land with pyramidal tents set in orderly fashion. Each tent accommodated four men, and it was immediately apparent that easy living and convenience would have to be forgotten for a while.

Schedules were posted, and the first weeks passed swiftly with orientations, class lectures, tactics, and field problems. Then later the rifle, pistol, machine gun, and mortar ranges were added to the program along with classes in map reading, ballistics, and leadership lectures. All were crammed into a tight schedule, and time seemed to fly.

The rest periods were always welcome. Some talked and others just laid down and rested, but those who talked always got around to their favorite subject, school activities for their, soon to come, senior year. Each one of their group were discussing some of the possibilities.

"I wonder if we'll have a big freshman class this year?"

"Well, it gets bigger every year and more girls each year, that's a help." They all laughed and another one commented briefly.

"Now maybe we can find some girls on campus, and we won't have to drive all the way to Columbus for a date."

"No, not for me, those girls at M.S.C.W. are too pretty to be left all alone."

"James, do you remember last year when Bill Taylor got that old county school bus and would take a load of students every weekend from State over to Columbus?"

"Yeah, we had some great times. And we'll have some good times this year with lots of trips out of state, 'cause that football team of ours is really going to win some games."

"I think 'Scabbard and Blade' is already scheduled to perform at the North Carolina game. But, you know what? I don't care if we lose every game, just as long as we beat those 'Ole Miss' Rebs."

"Boy, you can say that again."

The whole group started laughing just as a shrill whistle blew for the end of the break.

"Hey, gang, there's the whistle ... Break's over ... Back to the firing line."

It took a while to get everyone in their position and settled down for the firing. Then came the familiar yelling of commands, "Ready on the right ... Ready on the left ... Ready on the firing line ... Commence firing."

Week followed week, and soon a group of suntanned, bivouac weary cadet officers climbed aboard the trucks and rode back to the campus at Starkville. Then, they traveled home for a few days of rest.

The fall semester had started off with the expected surge of new students, and many classes were filled to capacity. This was good news to all concerned to see 'State' growing and on the move. The fraternities were especially elated. The large freshman class meant that most of them would fill their pledge quotas for the coming year, and the fraternity houses would be filled with new members to take the place of seniors who had graduated last year.

The Kappa Sigma Fraternity was one of the larger groups on the campus, and they, like others, would be seeking some of the best men in the new class for membership. Their meeting on that early autumn day was to install new officers. The incoming president and vice president had already been installed. The new president was speaking, "Our next office is Grand Master of Ceremonies. Will Quinn West come forward, please? Quinn, do you promise to faithfully perform the duties of Grand Master of Ceremonies to the best of your ability, to aid in keeping the ceremony and ritual of this fraternity in its true form, and to assist in the instruction of new members in these same ceremonies and rituals?"

"I will."

"Then by the authority vested in me as President of Kappa Sigma Fraternity, I hereby proclaim you as Grand Master of Ceremonies." Applause began to break out as Quinn returned to his seat. The men were proud of their fraternity, for its very foundation was based on Christian fellowship and brotherhood. Men like West reminded them that through the years, the fraternity had held fast to the goals of its inception.

Many of Quinn's best friends were Kappa Sigma, and one of the closest of these was Tom McCord. They had roomed together at Main Dormitory during their early college years, and they had very similar goals in life.

After the fraternity meeting ended, and the usual fellowship and congratulations exchanged, Tom and Quinn walked slowly back to their dormitory room at Hull Hall. They walked for a while in silence and Tom finally spoke, "Quinn, you know that you are really going to be carrying a heavy load this year with the leadership of the YMCA on your shoulders, as well as your ROTC and fraternity commitments."

"Yes, I know, but it won't be too bad. I'm going to lean on you a good deal for help in the 'Y' meetings, and the B.S.U. Council. Do you think you'll have time to help?"

"Sure, you know I'll help all I can. I enjoy the 'Y' work anyway, so we'll be able to get it done."

The schedule as President at the "Y" did prove to be demanding of Quinn's time, but it was a project that was near to his heart. The YMCA was one of the largest organizations on campus, with more than two hundred cabinet members and thousands of students who took part in its varied activities during the school year. The large three story YMCA building was filled each day with students who were there for study and recreation, as well as special programs or luncheon and banquet engagements. Quinn realized the importance of this organization for furnishing Christian fellowship, and inspiration for all students on the campus. Much hard work was done setting up committees, arranging for speakers, and coordinating these programs to tie in with other college activities. It was, in reality, the students' "home away from home," and they could always expect a warm friendliness about the place as well as a relaxed, informal atmosphere.

The fall weather felt good and with it came all the excitement of sports, especially football. The early predictions of a winning football team at State proved to be accurate for they didn't lose a single game during the entire year. The team went to the Orange Bowl in Miami to beat the powerful Georgetown "Hoyas," 14 to 7. Their schedule had been a tough one with Alabama, Ole Miss, Auburn, LSU, Millsaps, and North Carolina as some of their strongest opponents, but they came through with flying colors.

The "Scabbard and Blade," an expert marching group of Cadet Officers, had performed at several of the out of town games and had given some skillful demonstrations. In addition, there were the fine presentations of the "Maroon" marching band, which made a very exciting year for State football.

The Christmas holidays were swiftly approaching. As always, it put that extra measure of cheerfulness in the students and in the school surroundings. Christmas decorations were adorned on many of the buildings, and the "Y" had outdone itself that year.

A Christmas tree had been placed in the large main room, and students had decorated it with every bit of tinsel and paper cutouts imaginable. Other decorations were placed throughout the various rooms, and a warm glow settled over the students as they discussed plans for the coming holidays.

Quinn and a younger friend named Bill Taylor were talking about getting rides home for Christmas.

"Quinn, have you gotten a ride home for the holidays already? I'll be going your way with a bunch, and you are welcome to go with us."

"No, I think I'll stay a few days longer and teach my class at First Church."

"Well, they probably won't care if you miss just one Sunday, will they? And you'll have to hitch-hike home if you let everyone leave the campus before you go looking for a ride."

"I know, but I really like those kids, and they always seem to look for me each Sunday. I'd hate to disappoint them by not being there."

Bill spoke with admiration, "Quinn, you are an unusual guy, you know it?"

"What makes you say that?"

"Well, there aren't many guys who would spend their holidays teaching a bunch of kids."

"No, it's really not that way, because the thing is, I like my class, and it's not a big thing to give them some time."

"Okay, I understand. Anyway, I'm sure those kids must love you if you care that much about them. I'll see you and the gang after Christmas holidays."

Later that week Quinn made it back to his home with one of the church members going his direction. The holidays left him refreshed and ready to tackle his last semester of college.

The senior class was large that year with almost five hundred students, and it had been a difficult task for the honorary fraternities to select their few members out of so large a class. After the holidays were over, most of the honor students had been selected, and ceremonies of installation followed for the new members. Quinn had been selected for membership in O.D.K., Blue Key, and Scabbard and Blade. Each one of these organizations selected their members based on scholastic record, service to the school, and leadership ability. Other honors were to come to him during his senior year such as Hall of Fame, Who's Who in American Universities, and Gold Triangle, an award for outstanding work in the YMCA. However, all these honors and awards never affected Quinn's friendly, outgoing personality.

The weeks slipped by with increasing activity leading up to the culmination of another school year. Everyone was happy about the approach of summer vacation, especially the seniors with their completion of classwork and graduation not being far away. The graduation ceremonies were to be held at Scott Field that year because of the size of the outgoing class. So many parents, friends, and fellow students were expected, it would take a football stadium to seat them all. The program would be a long one with honors and awards scheduled to be given individually. Awarding of the Army Reserve Commissions to the sixty cadet officers were to be given as well.

The day finally arrived for the great occasion, and the stadium was filled with the expected crowd. Speeches were made, awards given, and the long line of graduates began to file by the podium as their names were called to receive their diplomas. It was a beautiful event.

The ceremonies had nearly come to the conclusion when Mr. Hilbun came to the microphone and presented one last award. He addressed the gathering, "Faculty, graduates, and friends ... There have been many honors given today by the selection of the faculty, but one award remains which is being given by the vote of your entire senior class. It is the 'Bertha M. Scales Award' given each year to the one senior whom you have selected as having led the most exemplary Christian life while here in college. It gives me great pleasure to announce the recipient of this award ... John Quinn West, Jr.!" Applause rang throughout the stadium as Quinn arose and tried to blink back the tears that had welled up in his eyes. He was so pleased with the award and so humbled that fellow students had recognized his life as being one of Christian dedication. Other awards would come in the years ahead, but none would touch him as deeply as had this one.

The receptions that followed were filled with the remembrances and fond farewells between the graduates. Parents were exchanging greetings of happiness at the thought of their children's accomplishments and professors were bidding goodbye to students they had grown to love over the years. It was a very happy occasion, but few of the people gathered here would ever suspect how bravely these young men would fight for their country in the coming years ahead. Even fewer still would realize that for many of these men, it was their last farewell. They were the Class of 1941, a year which would live in the memory of Americans for many, many years.

Aircraft "X2-V" crashed at Rivenhall during transition training. Left tire blew out on take-off collapsing the geaar and left propeller hit the pavement. This cartwheeled th engine completely off the left wing and plane burned to the ground. All crew escaped

CHAPTER FOUR

U.S. ARMY AIR FORCE PILOT TRAINING, 1942

The expected letter came from the War Department quickly after Quinn's graduation, and the message was brief and concise. It read, "The following Infantry Reserve Officers report for duty 0900, July 19, 1941, to the Commanding Officer, Fort Bragg, North Carolina." Quinn's name was among those listed. For most of these men it was a case of resignation to Army life; however, Quinn was elated.

For several years, he had been dating a very lovely, young lady from Ashville, North Carolina, named Ruby Hogan. They had first met at Ridgecrest, a scenic and inspiring summer church camp in North Carolina. He was immediately attracted to Ruby's youthful beauty and her idealistic goals in life that were so similar to his own.

Quinn was a handsome young man himself, and there were many girls he had dated in high school and college, but after Ruby came into his life, he immediately realized she was his choice for a marriage partner. Their engagement was announced soon after leaving Ridgecrest and now their wedding plans could be completed with his new Army assignment designated so near to her home.

Shortly after getting established at Fort Bragg, they set a date, August 28, for their wedding ceremony. It was planned as a small wedding with the family and a few close friends in attendance. The bride's father, Reverend Perry Crouch, officiated, and the ceremony was a touching, beautiful experience with many friends present at the wedding and the reception which followed.

Despite the nuptials, the work at Ft. Bragg continued. Every four months a new detachment of men for infantry training arrived. It was monotonous for Quinn to train these new men because each group were given the same courses and taught in the same way.

There were orientations for the new men, close order drill, and weapons familiarization. Then later they went through weapons range work, field problems, bivouac, and graduation. The men were then shipped out to permanent bases for duty, and the next group of recruits would arrive quickly following them for Basic Training.

West began to realize this type of duty wasn't what he really wanted to do in the Army. He was looking for a greater challenge, something that would give him freedom to exercise a skill, something that would put him in touch with machines. He was already missing the tractors, combines, and machinery of the farm back home. Since he had always wanted to fly an airplane, maybe with a little luck, he could get a transfer into the Army Air Cadet Program like some of the other officers had done recently. It required the approval of his present commanding officer and some exacting physical and mental tests, but he decided to apply.

Months passed with no approval, and as December approached, Quinn had enough time in service to be eligible for a Christmas leave. Everyone on the Base wanted to take the last two weeks in December, of course, but this was impossible. Some would have to go early and take the first two weeks in December, so that others might go later. Quinn chose the early segment, and the first week was a grand welcome home. He and his wife made the rounds to visit all their relatives and friends. He was happy to be home again, but more than this, he was excited about introducing his new wife, Ruby, to all the relatives.

The first week of visiting was warm and cheerful, and Quinn was in an exuberant mood. Their schedule even included a stop in Memphis, where Quinn could visit for a few hours with his aunt and uncle, and a kid first cousin whom he had taught to shoot a shotgun years ago at Sardis. When Jackie came in from school the afternoon of their arrival, Quinn gave him a warm smiling welcome, "Hey, Jackie, how is school going for you? Man alive, you have really grown since I saw you last. I can't believe you have gotten so tall." Jackie was grinning with pride to think that Quinn had come to see him. He liked Ruby immediately, but he was a little speechless around her. She was so pretty and so cultured, it was difficult for him to find the right words of conversation.

Quinn looked handsome in his officer's uniform. The army officers called it "Pinks and Greens," a beautiful combination of a forest green jacket and rose beige trousers. There has never been a better looking army officer's uniform before or after.

Quinn spoke again, "Jackie, you and your folks will have to come down to Sardis and see us again. We'll be here for another week before we go back to Ft. Bragg."

The short stay continued with much laughter and conversation about the "good ol' days." However, soon it was time for goodbyes and the family walked with Quinn and Ruby out to their car. Quinn was walking with his arm on Jackie's shoulder, and talking to him. "Well, take care of yourself, Jackie, and study hard in school. We'll be thinking about you."

Jackie wanted to say that he would miss him, but the words just didn't seem to come out. Instead it was just a simple, "Goodbye Quinn."

Quinn and Ruby drove back to Sardis that evening to join the family for an elaborate dinner, an early Christmas celebration in honor of his homecoming. It was Saturday evening, December 6, 1941. Unknown to them, it was the last evening Ruby and Quinn were to know a world at peace.

The next morning the family attended church. Quinn made the most of the precious time with his family. In the afternoon he and Ruby visited family friends. As they were getting in the car to go back to the farm, someone across the street told them about an important radio message.

Later, a special broadcast by President Roosevelt announced that the Japanese had bombed Pearl Harbor, December 7, 1941, "A day that would live in infamy." All military forces on leave were recalled to their bases, and all future leaves were cancelled. The nation was quickly getting on a wartime footing, and every branch of service was attempting to build its strength, as soon as possible.

The Army Air Force was opening every possible base for an enlarged cadet program because the shortage of pilots was crucial as well as the shortage of air crewmen, bombardiers, navigators, ground crewmen, and thousands of others who would be needed to fill the ranks of the vital air service. The conflicting urgency and sadness plagued Quinn as he packed to return to Ft. Bragg. He realized the impending responsibility that lay with America's military. He was proud to contribute, to protect his life gifts but there was still a part of him that wanted to delay leaving the comfort of his family and hometown.

Shortly after West's return to Ft. Bragg, his request for transfer was approved and orders came for his move to preflight training for testing, and then assignment to pilot training at Helena Air Force Base, Helena, Arkansas.

The base at Helena was one of the civilian air fields which had been quickly converted by the government. It was placed in the Southwestern Air Command to train pilots during the primary phase of cadet instruction. Many of the instructors were civilians, although they were well qualified for their military training duties at the air base.

Primary flight training consisted of many long hours of ground school and theory of flight before actually flying a plane. But finally, the phase of flight instruction arrived, the moment the cadets had been eagerly awaiting. As they attended classes, they could see the beautiful Stearman PT-17's lined upon the airport apron near the classroom windows. It was difficult to think of anything except flying when they saw the blue and yellow biplanes taking off and landing on the distant runways. They were required to log forty hours of flight time before graduating and transferring to another base for basic training. Those were some of the happiest flying hours of their careers as the instructors worked with them until they were ready to solo. The high point was when the instructor told them, "Okay, take her around by yourself ... You're ready to solo."

It was exhilarating to experience the freedom of flying alone, to feel the responsiveness of the aircraft as it banked and climbed. It was a new world, a new dimension, a giddy feeling of being in the air and flying free like an eagle. The cadets knew that it was the serious business of training for war, but it was difficult to feel anything but joy when a young, new pilot was climbing through a cloud bank into the pure blue of an open sky.

After the successful completion of Primary, Lt. West's orders sent him to Gunter Field, Alabama, for Basic Training. Gunter was an altogether different place than Helena Aero. It was a fully established military base, and the discipline was heavy. The officers and cadre at Gunter were Army Air Force men, and they took their job of molding Air Force pilots seriously. Requirements dictated seventy hours of flight time and very little horseplay. The cadets' flying became more like work, as their skills were sharpened to a fine edge. The flying included advanced maneuvers, formation flying, instrument and night flying, and the ever present ground school.

Their training aircraft was the dependable Vultee BT-13, called by the students the "Vultee Vibrator." It was a low wing monoplane with sleek lines and greater speed than the old biplanes. It looked much like a fighter aircraft, and it gave the cadets a sensation of what it was like to fly a real military aircraft.

The instructors and the classes were demanding, but for Quinn, the flying was sheer pleasure and excitement. It became routine in time, but there was always the satisfaction of increasing skills and realizing he could handle a complicated machine like the Army's fighter aircraft.

Many men couldn't meet the extra demands which were essential to become an Air Force pilot, and they were "washed out" of Cadets and reassigned to other Air Force programs. However, those who finished the phase at Gunter had passed the most difficult part of training. They were regarded, at this point, as nearly certain of completing their course of training and winning their silver wings of the Army Air Force.

The last segment of training was called Advanced Flight Training, and West was stationed at Blytheville Air Force Base in Blytheville, Arkansas, for this part of the work. The base at Blytheville was designed to train multiengine pilots to fly the medium and heavy bombers our country was manufacturing by the thousands to help in the war effort. Other schools would train the fighter pilots using the North American AT-6 for that purpose. However, the Blytheville School used the Vultee BT-13 during much of its training, and near graduation the pilots received their multiengine training in AT-9's and AT-10's.

The discipline was not as strict in Blytheville as it had been at Gunter Field. The cadets were treated with respect and trained as if they were commissioned officers already, rather than just cadets.

As graduation approached, the men began to speculate what assignments would come their way for further training in multiengined aircraft. Quinn's orders came after graduation for him to report to MacDill Field, Tampa, Florida, for training with the 21st Bomb Group in Martin B-26's, one of the fastest medium bombers in the world.

Other pilots were assigned to B-17 Fortresses, B-25 Mitchells, and B-24 Liberators. For Lt. West, the Marauder, as the B-26's were called, had a special appeal because he had heard it was fast and touchy to handle. It was the bomber that flew like a fighter, and Quinn was excited by the challenge. Other pilots had some bad experiences and were hesitant about flying the B-26. There had been some accidents because of propeller "runaway" and overloading, but most of these imperfections had been cured and the accident rate had leveled off. Still, the rumor about MacDill was, "One a day in Tampa Bay," and this was a little unsettling.

Quinn didn't mind the good natured joking from the others about his B-26 assignment. There was just something he liked about this highly maneuverable and super fast aircraft. Even after months of combat missions, he would give his opinion of it as, "The best ship in the E.T.O."

CHAPTER FIVE

MACDILL FIELD AND THE MARAUDERS

The Air Base at MacDill was at a high level of activity when Lt. West arrived in April, 1943. It was an operational training command, and as such, it was the collection point for the many groups of specialists who would eventually go into combat in the B-26. Pilots, navigators, bombardiers, air crew, ground crew, support teams, and technicians of every type would work together as a team at MacDill to finalize the last phases of training. They had been trained in their own specialties at hundreds of other Bases across the United States, and after graduation, they were sent here for the operational staging.

The pilots were sent here especially for transition flight school. It was difficult for a pilot to finish his Advanced Training in a light twin engine aircraft, then jump to the much larger, technically advanced, and fully instrumented Marauder.

For Quinn, it was a thrilling challenge; a new relationship between man and machine. Just to be able to tame the unpredictable intricacies of this medium bomber that had such a fearful reputation among Air Force pilots gave him deep satisfaction.

After checking in at Operations and finding his place at the Officer's Quarters, he looked over the class schedules. Orientation, B-26 systems and functions, hydraulics, electronics, maintenance, flight characteristics, fire and safety devices, and many other subjects were required to train the Marauder pilot.

It looked a little overwhelming at the moment, but it would all fall into place later. Orientation was scheduled for 0800 in the main theater building, so Quinn walked in that direction.

The theater building was large and filled with the hubbub of conversation between the new incoming pilots. Soon they were seated, and one of MacDill's Operations Officers began the briefing session.

"Men, you are going to be faced with a challenge in the months ahead, flying one of our country's most versatile weapons, the B-26. She's fast and maneuverable, and has more fire power than most heavy bombers. She'll perform for you if you learn everything about her from nose to tail, but she'll kill you, if you don't learn all her characteristics, attitudes, and loadings. That's what this transition is all about. We want you men to grow with this challenge. You will have twenty-five hours of familiarization of systems before even flying. Then you will have forty hours of flight time and later twenty hours of night flying, and that's just the beginning. Most of your instructors here have had two hundred hours or more in the Marauder. And believe you me, one hour of flight time in this aircraft is a long time because she requires your constant attention. It is no secret that you men are the top third of your flight school classes, and there is a reason for you being here. We need men who have exceptional flying skills, but as a bomber pilot, you will need organizational talent and ability to make rapid decisions in combat. These decisions will affect others in your crew. Their lives depend upon your excellence. We think you have got what it takes, so don't let us down."

Quinn felt the weight of responsibility tightening around him as he listened intently to the officer's lecture. He had not thought of his skill as a pilot becoming crucial to his crew's life or death in combat. It was a sobering thought, and it gave him a renewed urgency to be the best pilot he could be. Not so much for his own welfare, as it was for the lives of those men who would be flying with him in combat.

Schedules for the following weeks were passed out, and the orientation dismissed, as the men began to look around for familiar faces of friends from their respective flight schools. Quinn didn't see anyone from his advanced flight school, but he had met a few of the pilots already at his Officer's Quarters. Lt. Crummett, McLeod, Williams, and Ryherd were just a few names of new acquaintances now; however, later these men would fly beside him in combat. Their excellent flying ability would pull their squadron through many difficult situations.

The classes went swiftly with each day of learning the Marauder's parts and memorizing emergency procedures. Nearly an entire aircraft had been taken apart, piece by piece, and attached to recognition boards, so the men could identify each part and its function within each system. Flight procedures were memorized and became second nature. Instrumentation on the Marauder was considerably advanced for these new pilots, and each student had to understand all the panel's instruments completely.

Much of the flying at MacDill was routed over water, which necessitated training in emergency ditching procedures and practice drills. Rules of the airfield on landing and taking off, international rules of ocean flying and safety, study of navigational aids; visual and radio, as well as, celestial and instrument, all had to be studied and practiced. There was so much to be learned in so short a time, but finally after many weeks, the pilots began to eagerly look for their names on the flying schedules. They knew it was time to test and apply all the difficult studying.

Lt. West was to report to Captain Morrow at operations on the north apron at 0900. The day was a clear one and beautiful for flying. Captain Morrow began the conversation, as they started suiting-up for the flight.

"West, how do you think you'll like flying the Marauder?"

"Great, Sir, she looks like a fine aircraft."

"Well, you'll have to watch her every minute. One mistake can be hazardous, so keep your eyes open. One thing you'll notice right away is that speeds at take off and landing will seem excessive, but you'll get used to that. She stalls at 120 MPH, so watch that airspeed indicator, and give her plenty of time on take off. Landings are worse. You'll think you are coming in with too much power, but she settles fast, so come in at 135 MPH with a nose low attitude, and flare out just over the runway markers, so she settles on the mains, and let her ease down on the nose gear when the air speed drops down a little. Come on, let's look her over."

They walked out to the apron, and Captain Morrow continued his discourse, "Always make your preflight visual checks. You know the procedures. Train your men to have a healthy respect for those twelve foot props. Go on up the ladder and take the left seat. I'll fly the right to be close to the emergency systems."

Quinn climbed into his seat nervous and pleased at the same time. He was eager to take over but waited patiently for his instructor's direction.

Morrow continued, "Now you can begin your instrument checks. This first time I will talk you through most of the points even though you know these things already. Okay, while the men are hooking on the external power, continue checking your engine instruments and settings on the engine control levers. Now, pull the cowl flaps open, and check fuel gauges, set throttles at 3/4ths inch open, and turn on the master switch."

West handled the switches deftly, checking engine and fuel gauges as he went through the preparation for starting the left engine.

"Okay, go ahead and prime the left engine, now throw the energizer switch to the left and hold it for 30 seconds. Now throw the mesh toggle to the left, and ease the mixture to auto-rich."

The propeller began turning slowly as the engine coughed white exhaust smoke and began hitting on several cylinders then suddenly she was alive hitting on all cylinders and a smile broke across Quinn's face.

"There she goes, now ease the throttle forward, but keep her under 900 RPM for a minute. Now, repeat the sequence for the right and let both your engines warm up to about 1,000 RPM for a minute."

A feeling of exhilaration was sweeping over Quinn as he pushed both throttles forward to maintain 1,000 RPM. He could feel the surging power of the twin engines. It was a thrill he would never forget.

"Cut your battery switches on when the ground crew pulls the external power. Call the tower for taxi clearance, release the parking brake, ease the power on, and head out to the main runway."

Lt. West called the tower with an authoritative crispness, "Aircraft Number 312 requests taxi instructions ... Over."

Morrow began his instruction again, "After you reach the main runway, run your engines to 2,700 RPM and check the manifold pressure. Okay, let me take her this first time and watch what I am doing."

The plane began to accelerate and Captain Morrow continued his instructions, "Pull back until the nose wheel breaks ground then ease the control column to neutral." The Marauder was moving faster now with a slight nose up attitude showing signs of wanting to break ground and fly. "Let her pick up speed to about 135 MPH, and pull back on the column 'till she breaks from the runway."

West was experiencing that beautiful sensation feeling the Marauder lifting majestically with engines pulling at peak efficiency and the force of acceleration pushing him tightly against the seat's back.

"Pull the gears up, climb out to 500 feet before your first turn, and don't make any turns until you hit above 150 MPH. Turn her easy with no more than 15 degrees on a turn. You'll be making some wild turns in combat during evasive tactics, but now is not the time for that. Just keep her at about 170 MPH, and treat her nice and easy. Okay, you take over now and get a feel of the controls. Take her up to 2,000 feet, and make some slow turns."

Quinn could hardly wait to take the controls, and feel the ship respond to his maneuvering. Turning left, he pulled the wheel over for left aileron, a little left rudder with the pedals, and slowly pulled back for a little up elevator. Watching the altimeter, he saw that he had lost altitude, and he could feel her sinking in the turn, so he pulled more up elevator, but it was too late to hold a constant altitude.

"She really sinks fast in the turns, doesn't she?"

"Yes, you are going to have to feed in more elevator to keep her up. She is definitely a heavyweight, and those turns get real critical when you are close to the ground. She'll stall on you in a minute if you let her get too slow in a turn. Go ahead and try some shallow diving turns and watch how fast your airspeed builds. She'll redline on you very quickly if you aren't careful."

West was catching on fast and having a great time flying the Marauder. If this was work, he would be happy to work all day with this aircraft. It was one of the greatest thrills of his life to pilot the Marauder and feel it respond to his control. The surge of power from the engines was exhilarating, and he knew that he was flying an aircraft that required superb airmanship. It gave him a feeling of accomplishment to be able to wheel the B-26 around through her paces. Other pilots might be afraid of her, but for Quinn, it was a perfect blending of man and machine when he really began to learn how to fly her.

Because of his background on the farm, machines had always fascinated him. He could adjust and repair most mechanical parts; however, it wasn't the mechanical process that appealed to him most. It was the operation and capability of human skill which he liked best. His friends had always said he was an excellent automobile driver, but that he drove too fast for their liking. He would have probably made a great race car driver; pushing his racer to the limit on the straightaways, watching the turns, feeling every pulse of the motor, and getting all the efficiency he could get out of it.

A good race car driver, as well as a good pilot, is not just a man with a heavy foot on the accelerator or heavy hand on the throttle, but he is one who can nearly become part of the machine in the race. He is fine tuned to its capabilities, and understands each of its limitations.

Flying requires some special skills and abilities such as depth perception, timing, and balance or equilibrium. But, more than these, there should be an innate "feel" for flying, which some have, and others never seem to develop. This ability only surfaces in the actual piloting of an aircraft, so it is the instructor's job to train and watch for this skill in piloting. Some men feel comfortable at the controls and feel that the aircraft is an extension of themselves. They pilot with a control that is more automatic, rather than simply a thought process. In the beginning stages of flight instruction it is, of course, a thought process, but later it becomes automatic for some, and for others, they never seem to get the "hang" of it. So for these, flying continues to be a thought process, and slows their reaction time which makes for poor landings and uncoordinated maneuvers.

Captain Morrow was beginning to sense a special ability in Lt. West, as he saw him wheel the Marauder around with a grin of accomplishment on his face.

"Okay, Lieutenant, let's take her down to 500 feet, and get in the landing pattern. Call the tower for landing instructions, slow her down a little, and make a shallow turn into the final leg. I'll take her in this time to show you a few things." The Marauder began to turn onto final smoothly with Captain

Morrow at the controls.

Quinn watched his instructor closely, intent on learning every small detail as the ship began to approach the runway.

"Get the gears down, and flaps at 30 degrees, now slow her to about 135 MPH, and keep that nose down." The aircraft was in a slight nose down attitude as she quickly approached the runway. "All right, when she clears the outer markers, pull back and let her down on the mains, then slowly let her down on the nose wheel, and begin to apply the brakes."

It was a good landing. Fast but perfectly coordinated as the brakes began slowing the Marauder down to lower speeds and finally taxi speed before the taxi strip. "Looking good, you go ahead and taxi her in. Complete your instrument checks, then fill out the 1A forms and we're through."

West was fascinated by the speed and maneuverability of the Marauder. It felt right to him. He was comfortable with all the controls and the instrumentation. He knew he had made the right choice. It was just his type of aircraft.

"Well, was she too fast for you?"

"No, Sir, I enjoyed every minute of it," Quinn laughed.

Captain Morrow knew he meant it from the look and broad smile on Quinn's face.

"West, you will be working with several instructors here at MacDill. Some are easy going, some are very critical, but do your best and you'll make out okay."

"Thank you, Sir, I'll remember it."

Quinn checked in at the Operations Building after the flight, and then made his way toward the Officer's Quarters. Some of the men were waiting for his arrival. They had not flown yet and were eager to get Quinn's impressions.

"Well, how was it, West?"

"How did she handle on take off?"

"Was she responsive in turns and at slow speeds in landings?"

"How much did the instructor let you fly her?"

"Quinn, just start from the beginning and tell us your impressions of the whole flight."

"Everything turned out fine. The ship is beautiful to fly, but she's much faster and more responsive than I thought she would be. It's almost like flying a bullet."

"Okay now, don't try to scare us to death."

"We heard a rumor that all the instructors were tough as nails and would bawl you out for the least error."

"No, no, that's not true. The instructor I had was great. It was a good teaching situation. He explained everything and let me try the controls after we reached a safe altitude. Bergman is up now and he'll be coming in soon to give us his experiences."

Quinn was helping to relieve the tension and nervousness the others were experiencing waiting for their turn to fly.

He pulled a flight list from his jacket. "I think Taylor, Patterson, and Stangle are scheduled to fly tomorrow morning."

"Who's scheduled for tomorrow afternoon?"

"Probably Smith, Silverbach, and someone else, but I can't remember."

"West, aren't you scheduled for another flight tomorrow? Wonder what it will be?"

"Probably take offs and landings. We'll need plenty of that, because she's really a hot flying ship."

And so it went day after day; classes, dual instruction, instrument flying, cross countrys, night flying, formation flying, and night cross countrys. Gradually, week by week, the pilots were gaining proficiency in flying the B-26. Most importantly, along with this, the men were gaining confidence in the aircraft itself, and a sure feeling about their ability to handle her. They were becoming a proud bunch, even a bit cocky, knowing they were flying an aircraft which many other pilots were afraid to fly.

During the school, they were building valuable flying time, and some forged ahead with more hours than others. These few, after passing a series of tests, were classified as First Pilots. This meant that they would fly the "left seat" or be the senior pilot and aircraft commander. Others with not as many hours would be classified as copilots, and later become first pilots as vacancies came along.

Lt. West had passed the tests for first pilot and was promoted in rank to Captain about the same time. He was elated to be moving ahead in rank, but also it meant a better pay scale. His wife, Ruby, had found an apartment near him at MacDill, but suitable housing was so hard to find in Tampa. He thought about all the difficult times they had finding housing near the air bases. Ruby was so understanding about it all. Even with the discomfort of pregnancy she had followed him to every base without complaining. Now they were expecting their first child very soon, and he could do very little to help her with all his duties on the base. She was going to have to be strong and manage for herself. There were other pilots who were married and had their wives with them. One of Quinn's best friends was Dick George, who was at MacDill. The Georges were also expecting their first baby; so, the two couples developed a close and lasting friendship.

MacDill Air Force Base was always in transition with trained units beginning to move out to new bases or overseas, and other units moving in to be trained as operational units. The 394th Bombardment Group had been activated at MacDill and was assigned to a training base in Oklahoma. For a short time it seemed the base had lost half its men until new ones began to pour in from everywhere. It always meant a new unit was ready to be trained and forged into a fighting unit to be sent into combat.

Quinn had noticed the enlisted men's barracks filling again and longer chow lines at the mess halls. These were the men who would become a part of his own Unit, so the rumors said. New gunners, engineers, and radiomen, as well as officers who would become the Unit's navigators, bombardiers, and ordinance officers were all among the new faces. They were required to go through many of the same classes on the B-26 as the pilots since each crew member had to know each part and its function to work effectively as a team. They studied emergency procedures, operation and maintenance, as well as their own specific job in the crew. They worked hard at these classes, knowing that their lives, and the lives of their fellow crew members might hang in balance some day and depend upon their knowledge of the aircraft.

During the time the crew members were initiating their training in the classrooms, the pilots were building hours of flight time, honing flying skills to peak efficiency, and working together in formation and team maneuvers.

The framework of command was being organized also at this time. Incoming officers of higher rank, already familiar with the Marauder, were being selected for positions of leadership. In July of 1943, Lt. Colonel John Batjer became the first Commanding Officer of the Group. Immediately, he began carefully selecting his staff of officers for Group Headquarters. He also had to select other officers and men who were training in ordinance, supply, logistics, and operations, as well as, the Squadron Commanders of the flying echelon.

The Unit was slowly coming together and evolving into a self-sufficient command. There were medical teams, motor pool teams, and supply teams, all combining with other groups to make the integral Unit.

Orders began to filter down from Headquarters Command to the units and schedules were being made for the operational training of flight crew personnel. The pilots were prepared to train with the crews, and the flight crews were eager to get to the Marauders to begin training at their positions in the aircraft.

Capt. West picked up his first schedule at Operations and began to read over the orders slowly. "Crew assignment for aircraft #1282, Captain John Quinn West, Jr., pilot; Lt. R. T. Hughes, copilot; Lt. R. W. Hunsicker, navigator; enlisted crew, Sgt. T. J. Natanek, radioman; Sgt. H. Zola, engineer; Sgt. E. B. Picklesimer, gunner; and Sgt. Robinson, crew chief.

West had already had some flights with Lt. Hughes and Lt. Hunsicker and he liked both of them. Hughes was from Memphis, Tennessee, so they had interests in common during their conversations with home towns only fifty miles apart. West's home was just south of Memphis in Sardis, Mississippi, but he knew Memphis well from previous visits.

Hughes had worked with some of the enlisted men, so it was an easy way for Quinn to get some information from him about the new crewmen.

"Lieutenant Hughes, do you think this might be our permanent crew?"

"Yes, Sir, that's the rumor, unless you request a change. One of the guys I know in the 394th said there wasn't much change from the original assignments unless someone was dissatisfied and requested to be moved."

"Well, what do you think about the crew? Is Natanek a good man?"

"Yes, Sir, he's one of the best radiomen in the whole Group. He worked with radio in civilian life and he has been in the Air Force since before Pearl Harbor. He's older than the other crewmen; level headed, and should be a good stable influence on the other men. He knows that radio, all right."

"What about our engineer-gunner?"

"Zola? Well, he's young, but he is plenty smart. He's made some top grades in all his classes, and he looks as if he can take the '26 apart and put it back together again. He's a likeable guy too. Always friendly and joking. He'll be good for the crew."

"What about Sgt. Picklesimer?"

"He's a young guy too, but really smart. He knows his ordinance and has made high marks on the gunnery range. They say he can field-strip a 50 caliber machine gun blindfolded. I think he's a good man and a good addition to the crew."

"Okay, it won't be long until we get to greet these guys. You meet them at the hanger at 10:30 AM and help them get suited-up for the flight, and I'll take care of everything, and meet you at the plane in about 30 minutes."

Captain West went out to check over the aircraft before flight time. It was a habit with him now; checking control surfaces, checking instruments to see if someone had accidentally left a master-switch or ignition switch on. It took a while, but he had some time to spare.

It wasn't long before he saw a group of enlisted men walking toward the aircraft, suited-up with chutes and Mae West jackets.

They looked like they were all eager to get their first ride in the Marauder. They saluted as they moved in closer. West returned the salute and talked to them for a while.

"Which one of you is Sgt. Zola?"

"That's me, Sir."

Quinn sensed immediately that Zola was a good man. He was young and enthusiastic. His exuberance came through with his quick smile and ready answers.

"Where do you hail from Sergeant?"

"Dorchester, Massachusetts, Sir."

"I've heard some good reports on all of you men."

"Are you Sgt. Natanek?"

"Yes, Sir."

West could tell that Natanek had been in the Air Force for a while. He was restrained and showed a discipline borne of years in the Service. Yet he was the type that would be knowledgeable and dedicated with his assignments.

"They tell me you are whiz on the radio. Is that right?"

"I hope so, Sir."

"And you're Sgt. Picklesimer. A good man to have around when it comes to those fifty calibers. Right?"

"Yes, Sir."

Picklesimer looked young, but Quinn could sense an eagerness in him to do a good job. Probably the perfectionist type about his weapons, but not quite as talkative and outgoing as Zola.

"Okay, men, I'm not big on military discipline, but I am big on seeing a man give his job one hundred percent. Were Lt. Hughes and Lt. Hunsicker getting their gear on when you left?"

"Yes, Sir, they said they would be here in a minute."

"All right, here they come now, let's climb up, and get started."

The men swung up through the nose wheel opening and began to make their way to their stations. Lt. Hughes came into the flight deck and sat on the right as copilot. Lt. Hunsicker would be their navigator and his station was amidships, opposite Natanek's station at the radio.

By the time the men got settled and began their checking of equipment, Captain West had started the engines and was calling the tower for clearance to taxi.

"Number 1282 to tower. Request taxi clearance. Over."

"Tower to 1282. Use access strip number one to north end of runway four. Over."

"Roger, Wilco. Over."

The engines revved up smoothly as West wheeled his ship around and rolled out to the main runway.

"Lieutenant, we're going to make a fast take off, so get ready to pop those gears and flaps when we break ground."

"Yes, Sir."

Already they had reached the north end of the runway and were revving the engines up to 2,700 RPM and checking manifold pressure. It was an awesome sight and sound when those two Pratt and Whitney engines revved up to near maximum RPM. They were 2,000 horsepower each and when they began to turn, the prop wash was terrible.

West pulled the mixture control to auto-rich and pushed both throttles wide open. The plane began to accelerate quickly and the force began to push the pilot's bodies back into their seats. The noise was nearly deafening when the engines were running full throttle. The ship began hurtling down the long runway. West began to haul back on the control column.

In the back of the plane, Zola looked at Sgt. Picklesimer wide-eyed, as if to say, "If this is how the Marauder flies, man, what a ship."

The plane was continuing to climb, and suddenly she heeled over in a left turn and then just as quickly leveled off.

Captain West's voice came over the intercom, "Zola, how is she doing?"

"Fine, Sir."

"Lt. Hunsicker, double check your headings and wind drift, we're heading out over water."

Zola looked through an observation window and saw the ocean below. It sure looked different from this altitude, but it was really something to look out, and see the broad expanse of ocean below.

The aircraft continued to climb for a long time as West made a wide rectangular flight pattern over the Gulf. Then slowly the plane began losing altitude and gaining speed on a straight heading for a few minutes. Suddenly, there was a quick banking turn that put them in the landing pattern at MacDill, then another turn brought them on the final approach as the plane dropped quickly.

Zola and Picklesimer were watching through the plexiglas windows in the back, and it looked as if the plane was going to hit the water as it nosed down with both engines nearly going full bore. Suddenly the field appeared, and the plane seemed as if it would hit the ground before getting to the runway. Then, in a split second, the plane's attitude changed from nose down to nose up, and the screech of tires told them they were on the runway. A second later the nose wheel touched down, and they noticed the plane slowing as the brakes were applied while they turned into the nearest taxiway. In a few minutes they were back on the apron with engines stopped, and men piling out wondering if every Marauder flight would be that wild.

They would soon get used to the Marauder's wild ways. What seemed fast to them now would soon become commonplace.

West and Lt. Hughes stayed in the cockpit completing flight forms, as the men walked to the hanger to get out of their flight gear. The conversation between the men was animated with expressions of amazement while they all tried to talk.

Picklesimer exclaimed, "Whew, I would have sworn we were plowing into the drink on that landing until I heard those tires squeal when we touched down."

"Yeah, I thought so, too," Zola replied, "but, I really like this guy. Did you notice how he talked to us when we first met him? He might be a 'Hot Pilot,' but he still seems to have an interest in his crew."

"Well, I guess most of the Marauder pilots have confidence and are sure like him because they are trained that way. From what I've seen in the Air Force, it takes a hot pilot to fly these ships, and these guys are proud and walking tall when they really learn to handle them."

Zola answered, "I've discovered one thing about him. Don't let that southern drawl of his fool you, because when he is in a Marauder he can fly it faster than greased lightning."

"Yeah, I'll tell you something else. If I have got to fight this war in a B-26, I'd pick Captain West to fly it for us."

"Yep, you said a mouthful that time, Zola."

The men started laughing and began walking back to their barracks. They had placed their stamp of approval on the soft-spoken Captain from the deep south. They all agreed that he was a "Hot Pilot," but they liked the way he flew.

None of those men could have ever guessed at that moment that they all three Zola, Natanek, and Picklesimer as well as Sgt. Robinson, their crew chief would go nearly through the whole war with him. They were confident in placing their trust in his expert flying ability, and in times to come, he never failed to live up to that trust.

The flight crews were gathering confidence in their aircraft, their leaders, and themselves, as week by week they flew together. Formation flying and cross country flying were the order of the day, as each crew member worked in his position in the plane, simulating the time when combat would come.

The crews changed from time to time in other aircraft, but Zola and the others were still a part of Quinn's crew. Captain West was confident in each one of his crew members, for they had all lived up to the high standards he set for them.

The Group had nearly finished their training when some exciting news was causing rumors all over the Base. Zola came in first with it.

"Hey, do you guys know they are going to shoot a movie film on the Base?"

"Really? Where did you hear that?"

"I didn't hear it. I saw it in one of the hangers down the line. They are painting some of the B-26's with Japanese insignia, and the men say they are going to be used in a movie filmed right here at MacDill."

"Dog-gone, that will really be great, won't it? How can we find out for sure?"

"We've got a flight this afternoon. Let's ask Captain West."

"Sounds like a good idea to me."

Suspense was thick until they could get out to the field and ask West about the rumor. After chow at noon, they found him out on the apron talking with the crew chief, Robbie Robinson.

"Captain, we want to ask you about something."

"Okay, what is it?"

"Well, we saw some B-26's they were putting Japanese insignia on, and they said the planes would be in a movie."

"Zola, you are the best of the bunch at digging up rumors. I just barely knew about this one myself, but it's true. They are going to shoot a film down here. Major Dempster is in charge of it all, and Lt. Gleis is the Liaison Officer for the Hollywood group. Not too many of our officers will be involved in it except Gleis, Borr, Crabtree, Kale, Cordell, and ourselves."

"Hey, are we really going to be in it?"

"Yes, I talked to Major Dempster this morning. It's just going to be some fly bys with the B-26's in formation, and Gleis' ship fitted with a smoke canister to pour out the smoke when some fighter pilots attack the Marauders."

"But why are they using 'Jap' insignia on our planes?"

"We are the bad guys in this film. They say our planes look similar to the Japanese 'Betty' Bombers. The good guys are the fighter aircraft, and they will fly out of the Air Base in North Tampa."

"Will you take all the crew with you?"

"Sure, all of you can go. It won't be much different from our regular formation training flights, except just at the time the camera plane gets in place and starts shooting film, the U.S. fighter aircraft will come in close and act like they are attacking the bomber squadron. It will be good training for you guys to try to track those fighters with your guns when they come in close. We will be in radio contact with all the fighter and camera planes so when they give the signal, Lt. Gleis and Major Dempster will ignite their smoke canisters, begin losing altitude and appear to have been hit looking as if they are on fire."

"What if the Coast Guard Batteries start firing at our Japanese marked bombers?" Quinn smiled at his crews apprehension at being Hollywood villains.

"That's taken care of. All the Southwest Coastal Defense Units have been alerted that Japanese marked American aircraft will be flying in their area, and that they are not to open fire."

"Well, that sounds okay. When are they going to start filming?"

"Tomorrow, if the weather and cloud formation is right. We'll just fly our regular training schedules each day, and when they think conditions are right, they will let us know."

The men were excited and could hardly talk about anything else in the barracks that night except the filming. They wondered how much of the movie would be taken of the Marauder battle scenes, and they tried to guess who the Hollywood actors might be. Also, there was much discussion about whether the actors would come to MacDill for the filming. They would all just have to wait and see.

The following morning was dull with heavy cloud cover, and the men were disappointed until they found out that this was the type of weather the camera crews were looking for. Scattered cloud cover would give them their best shots. Bright sun and a cloudless day was not what they were looking for at all.

The preparations were time consuming, and it was nearly noon before the group had taken off and formed up at 5,000 feet to fly over some of the scattered islands in the Gulf area. The aircraft droned on, and the men were eagerly watching, trying to see fighters and camera planes. Major Dempster was leading the squadron with Gleis and West close behind. Radios had been set on proper frequency, and suddenly Maj. Dempster's voice began alerting his pilots.

"Blue Leader to squadron, the camera aircraft are moving in at 10 o'clock high. Maintain altitude, and prepare for mock battle."

West called over his aircraft's intercom, "Captain to crew, Battle Stations alert."

"Blue Leader to squadron. Fighters coming in at two o'clock high."

"Blue Leader to Gleis. Ignite your smoke canister on the first pass. I'll wait until they come around again and ignite mine."

"Gleis to Blue Leader, Wilco. Over."

The excitement was building by the minute, and the crew in West's ship were beginning to come in over their intercom.

"Here they come, Captain. Man!, look at those fighters peel out of formation. There must be a dozen of them."

"Sergeant, can you track 'em from the top turret?"

"Yes, Sir, I'm blasting away at them now."

"There goes the canister on Gleis' ship. Hey, look at him pitch that nose down in a shallow dive. It really looks like he took a direct hit and started burning."

"Yeah, he's breaking formation and spiralling down. Boy, that will make some good film footage."

"Zola, watch 'em. They're coming around again from behind at six o'clock high. Try and track 'em on those fifties."

"Okay, Capt'n, there's three of them on our tail now."

It was simulated combat, but it was as close to the real thing as one could get in training without firing live ammunition. Major Dempster had ignited his smoke canister and began to slowly spiral his ship down out of formation. It was a beautiful scene of mock combat, and the crews were thrilled to see such an aerial display.

That night at chow time the topic of conversation was the combat filming, and the crews' comments were loud and enthusiastic.

"Boy, did you see Lt. Gleis dive his plane down over the Bay area with smoke pouring out of that thing? It sure looked like the real thing to me."

"Yeah, I really liked seeing those fighters coming in too. When they would break and head for the ship it seemed so realistic."

"Has anybody found out what this movie is all about?"

"One of the other crews said they thought the actors were Spencer Tracy, Irene Dunne, and Van Johnson. They said the movie was called, "A Guy Named Joe."

"Boy, I wonder if we'll get to see any of those movie stars?"

"No, no such luck, all their scenes were shot in Hollywood, and these Marauder films will just be spliced in."

"Well, that's Hollywood for you. They'll be finished and gone tomorrow."

"No, I don't think so. Captain West said that they would be here for a while, and we would get some more training trying to keep our sights on those fighters."

The movie crew did stay for a while, nearly a month. There was one problem after another. The first sequence was spoiled because the camera plane cast a propeller shadow which was visible in the finished shots. Other times the smoke canisters failed to ignite at the proper time, and sometimes for days on end the filming couldn't be done because of poor weather conditions. But, in the end it was worth it all for the pilots when M.G.M. and the camera crews threw a gigantic farewell party at the Tampa Terrace Hotel. Wives and girlfriends of the pilots were invited and the partying continued until the late hours of the night.

The training was fast drawing to a close at MacDill as the men were getting proficient in their jobs. All the crews had become a team and knew they could depend on one another. Orders had already been cut naming them the 397th Bombardment Group, and the mantle of command had been placed on the shoulders of a tall Texan named Colonel Richard T. Coiner. He was a tough disciplinarian, but outgoing and well liked by his men.

Only one other phase of training was left in the operational training process, and that was to prove these men were steady under near combat conditions. It was important for them to train under living conditions similar to those which they would face at their overseas stations. This was the Air Force bivouac, and it was to take place at the Avon Park Bombing Range, an isolated area in Central Florida.

The main purpose of this range was to give the Group simulated bombing runs with live bombs. It would train them to fly formations which would give the best bombing patterns. More importantly, it was training for the navigators and bombardiers on getting to the targets, locating targets through the bombsights, and hitting targets from high and low altitudes. It was seemingly endless days of bombing training during the daylight hours, and nights of fighting mosquitoes and living in pup tents. It was the "K" ration, "C" rations, and an occasional quick, hot meal from the field mess kits. It was sand and palmetto bushes, tall thin pines, and it was hot.

The Florida heat was humid and sticky, and ground crews fussed and fumed with maintenance and bomb loads, but it was training, and the men were learning to work under difficult field conditions. The only good part about it for the crews was that it was only a few weeks long, and soon they would be getting bag and baggage together to go back to MacDill.

The Air Base looked beautiful to the men upon their return, and the barracks seemed like home after the grueling work schedules of Avon Park. Many of the air and ground crews got leaves to go home, while others got weekend passes to celebrate the completion of their training, But, soon it was back to the work schedules and back to the serious business of winning a war. There were many final jobs to accomplish before the Group moved out for overseas operations.

The aircrews were looking forward with great expectancy to the task of getting new aircraft at Hunter Field in Savannah, Georgia. They had been training for all these months in the old, olive drab painted aircraft which were considerably aged and bent from years of usage by other Marauder Groups who had trained previously in them. Now the time had come for the crews to move to Savannah to get a brand new aircraft.

One morning after chow the crews loaded into several C-47's and took off. The big, cumbersome Air Force transports droned slowly almost putting the men to sleep. They had left MacDill at dawn and the anticipation of new planes at Hunter had precluded much sleep the night before. They were like kids with a new toy when they first saw those silver Marauders lined on the apron at Hunter field. It seemed like hours instead of minutes when the transports landed and all of them sorted their gear while the pilots reported directly to the flight office.

Once the pilots were assembled, the officers in charge of supply and disbursement at Hunter began reading off the pilots' assignments to their aircraft.

"Captain West?"

"Yes, Sir."

"Your crew is assigned to aircraft number 296138. It's over on the far side of the apron."

"Lieutenant Smith?"

"Yes, Sir."

"Your crew is assigned to aircraft number 296139. It's also over on the far side of the apron next to Captain West's assignment."

Quinn walked out and signaled for his crew to come along.

"She's the first one on the apron men, down toward the end."

"Dog-gone, isn't she beautiful, Captain? New engines, new tires, new everything. I haven't been this excited since I bought my first car."

"Yeah, they're beautiful all right. A new shining package from the Martin Plant in Baltimore. Lt. Smith is assigned to the sister ship, just one serial number different. Isn't that something? These two babies went down the assembly line side by side."

"That aluminum finish looks like a sparkling diamond compared to those olive drab buggies we flew in training."

"When can we take her up, Captain?"

"Well, it will probably be tomorrow. Lt. Hunsicker will want to check and set the magnetic and gyro compasses, and then make his magnetic deviation card. While he's doing that, we can check the controls and instruments over carefully. Also, Robbie will want to see those engines running several times while we check cylinder head and oil temps, as well as manifold pressures. It's got to be right before we sign our acceptance papers. We'll more than likely fly here for a few days before we're ready for more formation flying and bombing practice at MacDill."

Methodically, the crew spent the afternoon completing their individual inspections. They checked out their Marauder inside and out. Even after dark, lights were turned on and off while instruments used in night flying were tested.

The next morning everyone rushed through breakfast and hurried to their assigned plane, glistening in the early morning sun. She almost seemed to be alive sitting there waiting and saying, "C'mon boys, let's get up there. I thought you'd never get here."

The men climbed aboard and scented the aroma of new wiring, new components and new engine smells that was to them the perfume of a brand new aircraft. In a matter of minutes the engines started quickly and easily. Soon they were in the air enjoying the luxury of a new ship.

The new plane flew like a dream. No fouled plugs, no missing or sputtering engines on taxi runs, no overheated engines, no hydraulic leaks, even the radios seemed to have less interference than the old planes. The men were pleased with her, and so was Captain West.

The Group knew they were ready for overseas movement and combat, but the higher echelons of command had not as yet given the orders. So they continued to train in formation and cross country flying while their skills got sharper and sharper.

Colonel Coiner notified them that they had been awarded a Group trophy from the Air Force Command for their record of excellence in training at MacDill and Avon Park. The work was easing off and the men's spirits were high so weekend passes and parties became more frequent.

The 397th Bomb Group was asked to put on several "air shows" over the eastern states. There were shows in Atterbury, Ft. Knox, Fort Benning, Ft. Sill, and Sheppard Field. The men were always enthusiastic in showing their outfit's flying abilities at other bases. They were a proud bunch, the cream of the crop, and they delighted in it. Wherever they flew there were a few "low passes" made over towns and air bases, and many so called "Emergency Landings" were made at airports so a crew member could make a quick telephone call to wife or family.

The most memorable stop over was made at Reading, Pennsylvania, where the Group had gotten weathered in for several weeks. The men had partied all over town before they left. They had really become endeared to the towns people during those weeks of bad weather, and so at take off time there was a large gathering of people to see them off. The Squadron's goodbye salute was a low pass over the field and downtown district. One of

West's crewmen, Sgt. Natanek, commented on this maneuver by saying, "If we had gone any lower, we would have had to obey the red and green traffic signals." From that time on, the Group's stay in Pennsylvania was known as the "Battle of Reading."

By November of '43, the entire Group had been moved to Hunter Field where checking and final adjustments in personnel were made. West's copilot, Lt. Hughes, had been moved to Lt. Richardson's crew, and his replacement was Lt. William Budge. He was a tall outgoing type of individual, and the men who had a sixth sense about personalities seemed to like him immediately. They quickly tagged him with the nickname of "Wild Bill," and he was easy going enough that he didn't mind the new name.

The next few months were scheduled for advanced training efforts at Hunter Field. The integrated fighting unit began to be welded together and emerge as a combat ready team. During February of 1944, the following orders were posted on the Operations bulletin board.

"By order of the Commanding General, the 397th Bombardment Group, Colonel Richard T. Coiner, commanding, shall proceed to Morrison Field, West Palm Beach, Florida, for movement to overseas duty. Ground echelons will proceed to Camp Kilmer, New Jersey, for overseas movement by Troopship."

Crew of "Lucky 7"; Lt. Blomberg, Lt. Short, Lt. George, Capt. Taylor, Sgts. Everett, Brazier, London and Russell.

CHAPTER SIX

THE LONG FLIGHT OVERSEAS

This was it. What they had trained so long for had now arrived, and the bustle of activity was heightening each day. Packing, organizing, double checking lists of equipment, then moving equipment by truck, rail, and air to depots for overseas shipping. It was a constant scene of activity and movement.

A farewell banquet and party had been arranged for the officers of the air echelon and their wives or girl friends. It was a big affair, and the men had been looking forward to it for days. Ruby had also been looking forward to the occasion and the chance to meet some of the wives of Quinn's fellow officers. She was facing the time with mixed emotions as she was happy knowing that Quinn and his group had completed the arduous training which had been so challenging for them. But she was heartsick and frightened at the nearness of their separation when he must leave for overseas. They had often talked about it. Johnny was now one year old and Quinn wanted she and Johnny to return to Ashville to be near her parents. She thought it best also, but how could she ever manage without his strength to lean upon. He had been a tower of strength for her throughout their married life and now he was leaving. Tears began streaming down her face at the thought of losing him. But she must be strong and not give in to her fears. She knew she must appear confident for his sake if not for her own. Yes, she would go to the farewell party and be confident for him. As the time arrived, they all gathered together sharing much laughter, joking, and good fellowship. The Staff Officers eventually called the group to order, and asked that all be seated. Several of the Squadron Officers and Colonel Coiner had some congratulatory speeches to make preceding the banquet.

Quinn had asked Colonel Coiner several days before the banquet if it would be permissible for prayer to be said as a part of the farewell activities. He hadn't answered, but he indicated that it would be considered.

At the conclusion of Col. Coiner's speech, Quinn was a little surprised to hear him say, "We would like to call on Captain West for prayer."

Quinn arose and began slowly, "Our most gracious Heavenly Father, we praise you for your many blessings to us, and we pray for your continued watchcare, as we enter our overseas service. Forgive us of sin, and be merciful to us, Dear Lord, our strength, and our Redeemer. Amen." There was a fleeting silence that settled over the room. It was no longer than the blink of an eye but in that slight pause after Quinn's prayer a common realization ran through the minds of all who were gathered there. In the days, weeks, and months to come there would be many times, more than they cared to think about, when there would be nothing to save their lives during seemingly impossible odds except a power stronger than the cruelties of war here on earth.

With the serving of food, the hilarity and laughter began again. It was a high time for everyone, even though it marked the closing of their stateside duty. They were looking forward to the new experiences awaiting them at other air fields and other bases.

The air crew's move to Morrison Field was accomplished quickly while they left the ground crews with the responsibility of packing and moving the Group's equipment by ship to their new overseas base.

At Morrison it was the same old story of hurry up and wait, but the teams continued formation flying and other aspects of training which would be helpful in the months to come. It was pleasurable duty for a while as West Palm Beach was the center of recreational activities for those who sought the sun and sea on the Florida beaches.

If the men were restless and curious about when they would move overseas, they soon would know some answers. Col. Coiner called an officer's meeting for 1030 hours one morning. The breakfast meal had been leisurely until after the announcement of the meeting. Naturally, the whole place became a mass of rumors, guesses, and observations about where the destination of the 397th would be. Most had guessed England, of course, because the MacDill trained units had all been sent there previously. No one knew for sure, but hopefully, Col. Coiner would give them the word at the meeting.

The room was filled to near capacity with pilots and officer flight personnel as the Executive Officer called the men to attention. Col. Coiner walked to the front and addressed the officers.

"At ease, men. I can't tell you exactly where our destination will be as that is classified information. But, we will be flying the South Atlantic route overseas, and you first pilots will be carrying sealed orders to be opened when the flight is underway. Tomorrow at 0700 hours we take off on the first leg of our journey. This will be one of the easiest segments and the shortest hop. Our first jump is to Boringuen Field, Puerto Rico. You will be briefed at each station before take off as to weather, radio homing frequencies, emergency frequencies, difficulties, and special condition of flying each segment. You have been working and training these past few months in navigational problems related to ocean flying to prepare you for this journey. It's not going to be easy, but I have the utmost confidence in your ability to get the job done. There will be no margin for error. Any error you make will mean that you will miss the designated airfield and be forced to ditch in the ocean. This is serious business, and I want you men to handle it in exactly that manner. We will each take off at one minute intervals, and we will not attempt to fly formation. Other groups have tried this plan in the past and found that it was too tiring for the pilots, and too dangerous when flying into heavy cloud cover. Each crew will calculate their own dead reckoning, and all of us will assemble at each new base when we land. Major Berkenkamp will lead the 599th Squadron, Major Allen the 598th, Major Wood the 597th, and Major McLeod the 596th. Your Squadron Commanders may elect to break the squadrons into smaller units and fly loose formations to aid in navigation. Our Group S-2 Officer will give the weather briefing for tomorrow's flight." Captain Wood stepped up to the front.

"Men, our weather looks good for tomorrow; widely scattered cumulus, and clear visibility at all altitudes. We will fly at 5,000 feet. You have been given all your headings, beacon signals, emergency frequencies, and navigational maps. Are there any questions? No passes for West Palm Beach tonight. Get a good night's rest and be up at 0600 for chow. We take off at 0700. Group dismissed."

A steady hubbub of conversation began as the men started to talk about the new assignment. Was it England? Most thought that it probably was England and by way of the South Atlantic.

The men all knew the challenge ahead of them. Ocean flying was dangerous, to say the least. The past few years, it had only been safely achieved by trained airline pilots who had years of experience in crossing the Atlantic. Now it must be done by a group of young Air Force pilots who did not have that wealth of experience, but who had a brash confidence in themselves to know they could do it. The significant difference was their eagerness to get into the air battles overseas for the sake of their homeland.

Telephone lines had been in use all through the night. Men were saying farewells to families, wives, or loved ones before heading into the combat zone. Quinn had talked to his wife that night, and it seemed that they were closer than ever now, since both had become resigned to knowing that the time of separation had come. Sometimes it is difficult to express just the right feelings which are in one's heart at a time of separation. Words cannot fully convey all of the love which is there. But Quinn and Ruby knew their love was strong and could weather the long separation. Their marriage was an ideal one, that was security and consolation to both of them. It gave them strength to face the long nights of loneliness and distance. It gave them a determination to keep their love strong and abiding, even though the world's political upheaval and war had required them to be apart.

The meteorologist's forecast was right as the dawn had broken fair. The weather was clear with a few scattered cumulus moving in from the ocean. The activity was more extensive than usual with crew chiefs checking over their aircraft. The officers carefully reviewed their check lists while gasoline trucks moved down the line to "top off" fuel tanks just before take off.

It was a historic moment; the culmination of months of preparation and training as the 397th Bomb Group began their long awaited exodus to their airbase overseas for combat operations against the enemy. Excitement was running high as the ground crewmen were hooking in the external power boosters and the inertia starters were beginning their low pitched whine to start the engines. First coughing exhausts, then firing with a roar, the engine noise of the Groups' aircraft rose to a deafening crescendo. Some of the lead planes were revving up their engines and starting to taxi out on the access strip moving toward the main runway.

Their Commander, Colonel Coiner, was leading the long line for take off. He was always a disciplined, stalwart, and courageous leader, tough on the outside, but understanding and considerate when his men were in trouble and needed his help. They admired him greatly, and they would follow him into combat without question or complaint.

Quinn and his copilot were going through their preflight check before taxiing out with their squadron. Their Squadron Commander, Major Allen, was ready to roll onto the taxi strip while Lt. Budge continued to check his instruments.

"Cylinder head and oil temp, okay. Hydraulic pressure, good. Set oil cooler shutters, altimeter to station pressure, and carburetor air setting, okay." Budge continued to rattle off the check list items unconsciously. "We follow Captain Bronson on roll out, don't we?"

"Yes, it looks like the line up is Allen, McLeod, Bronson, and then our ship. Are the ground crewmen clear on your side?"

"Yes, Sir, they're backing off now ... Captain, there's a woman holding a baby and waving near the edge of the hanger apron. She sure looks like your wife."

"By golly, I'll bet that's her. This morning when we said our goodbyes, she told me to look for a surprise later at the Field. She must have begged

Col. Coiner for days to get clearance to come on the Field ... That's her all right, waving that handkerchief for dear life ... She's looking straight at us."

"Do you think she can tell our plane from all the others?"

"Sure, she knows the serial number on our rudder."

"Well, I'll say one thing for her, she sure had to work a miracle to get on the Field for this operation. Are we ready to taxi out, Captain?"

"Okay, call the tower and confirm taxi instructions. Bronson is beginning to move out now."

West eased forward on the throttle controls and his plane moved forward, swinging onto the access runway behind the others in single file. They all moved slowly toward the main runway for take off. One by one as the aircraft reached the long main runway they stopped and angled facing it to turn both engines to 2,700 RPM. After a minute or so they would then release the brakes, turn into the center of the runway and give her full throttle.

Quinn could see the hangers and buildings surrounding the large concrete aprons, now in the distance, as he turned onto the runway. His eyes searched for a moment, straining to notice if he could see the wife he loved so much, but it was hopeless. The distance was too far. He pictured her, probably even now, still waving goodbye, and it warmed his heart to know that she had come to see him this one last time. He blinked back the tears which had welled in his eyes as he wondered if he would ever see her and his son again. "Yes, perhaps ... Sometime later," he was sure he would see them again.

The engines roared to full throttle, and the aircraft began to move ahead, slowly at first, then picking up speed faster and faster until the runway markers began to blur on each side. One hundred miles per hour, then 120. The nose wheel had already lifted, and she was getting light on the main gears. Now, 130, 135 and she had eaten nearly a mile of runway, but now she was lifting, gaining altitude slowly, and building flying speed for the first turn.

"Bill, pull the gears up, and set the rudder and elevator trim ... Hunsicker, how is it going in the navigation department?"

"Fine, Sir, we're on course for Boringuen Field, Puerto Rico. About five hours of flying time should put us there right on the nose."

"Robbie, how does the ship sound from your position?"

"Okay, Sir, she's looking good."

"Men, take a last look back at the good old U.S.A. That's the homeland we'll be fighting for in this war. God bless her."

The crewmen looked back to see the shoreline of Florida, slowly slipping away. They were all reflecting on personal thoughts, while a quietness settled over each one as they looked and thought about home. Their thoughts went back to loved ones, friends, and events of the past. A sweeping feeling of nostalgia and remembrances began to flow like waves over their minds, as they relived last goodbyes and fond farewells to those at home.

They had good reason to be nostalgic, and wonder if they would ever return home again. Like the rest of the war combatants many of the 397th Bomb Group would never set foot on the soil of their homeland again. For many of these men, it was a last farewell and a longing look at the land of their birth.

After an hour's flying time the flight had been near perfect, with no problems at all during take off and climb out. The engines sounded smooth and responsive to each change of throttle, mixture, and pitch control. She just felt like a good solid ship, and Quinn was in high spirits testing her response to each control change.

"Robbie, look off to your left, there's Major Allen and Bronson flying a loose formation with us. Those planes look beautiful in flight, don't they?"

"Yes, Sir, they sure do."

"Hunsicker, are you going to put us there by dead reckoning or radio compass?"

"Capt'n, this short hop will be some good proving ground for my dead reckoning. If we don't make it here, we'll never make the long hops."

"Okay, sounds good ... Men, go ahead and settle in. We've got about 672 more miles of ocean between us and Puerto Rico."

It wasn't long before the steady drone of the engines became nearly hypnotic. Not only that, but the endless stretches of ocean didn't provide much interesting scenery, so the men tried to stay busy with whatever they could do to keep from boredom or drowsiness. Hunsicker stayed busy rechecking headings and taking drift readings to confirm the ship's position calculating the estimated time of arrival. Robbie was kept busy checking the entire aircraft and its systems, and Natanek stayed at his post in the radio section. Sometimes they would look for other aircraft in the Group, and other times they would be talking to the other members of the crew. Captain West and Lt. Budge managed to combat boredom on the flight deck by taking turns every now and then at the wheel and relaxing for a while.

"Hunsicker, how is she looking now?"

"Captain, we're making about 200 MPH ground speed. If the wind doesn't make a change our E.T.A. at Boringuen should be about 1205. We have about an hour more of flying time."

"Okay, I'm about ready to start looking for the mess hall. Aren't you?"

"Yes, Sir, I could use a good plate of chow right now."

It wasn't long before West began to make a slow let down to 3,000 feet. Then a little later, they began to look for the land mass of Puerto Rico.

"Hunsicker, give me a present ETA."

"Sir, if my calculations are right, we are about ten minutes out from the airfield."

"Give us a quick check on the radio compass, and see if we are on course."

"Yes, Sir, she's swinging dead ahead."

"Bill, we should be seeing something besides ocean on the horizon by now."

"Every now and then I think I have spotted it, and it turns out to be only a cloud shadow on the ocean surface ... Hey, there it is, Capt'n, dead ahead."

"Yep, you're right. Hey Hunsicker, you've put us right on target. The airfield is on the northwest tip of the island, so we should be coming straight into it. There she is. Bill, give the tower a call for landing clearance."

"Looks like there are some other Marauders on the apron down there, so we aren't the first ones here, Bill."

The plane slowly banked around into the pattern, and then turned onto the final leg, and seconds later touched down with the familiar squeal of tires on pavement, as she settled in on the edge of the runway.

Quinn taxied down to the apron, and turned the ship slowly as she came into line beside the others parked there, then shut off the engines.

"All right men, we'll see you early in the morning after briefing. Don't get into any trouble. Just hit the sack early, and get a good night's rest."

The men found a truck waiting to take them to their quarters, and the officers loaded into a jeep waiting to take them to the Operations Building. Quinn, Bill, and Hunsicker checked in at Operations with the desk sergeant.

"Sergeant, did Col. Coiner leave any instructions for the 397th Group?"

"Yes, Sir, the officers are to meet here in Operations at 0600 tomorrow for briefing, and no one is to leave the post tonight. Corporal Williams will take all of you to the officer's mess and get your assignments to quarters for you."

Quinn was tired, as the long flight had been a strain. Not just the distance so much, but the worry about navigational errors, and the uncertainty of flying a new ship. Anything could happen in an unproven aircraft. Motors could quit, fuel lines could clog and fail to function, hydraulic systems could fail, and the complicated auto pitch props could go haywire. It had happened before to others, and it was the Captain's responsibility to meet those emergencies with measures to keep the ship and his crew safe.

After chow, Quinn headed to the bachelor officer quarters. A few others had gone to the post exchange and a movie, but he wanted to write a few letters home and get to bed early. The morning briefing would come all too soon, and he would be faced with another long segment of ocean flying. The barracks were quiet as Quinn walked in. Most of the officers would probably be returning from the PX or the officer's club later. He removed his jacket and lay back on the bed a few minutes thinking of Ruby and Johnny. He had already begun to miss them. He thought of Ruby even now as she would be packing her and Johnny's clothes and getting ready to make the train trip back to Ashville. He knew she would be happier there near her parents.

It wasn't but a few moments until the stress of the long flight began to catch up with him and he relaxed to close his eyes for a minute and sleep began to take over.

The next morning it was still dark when Quinn and the other officers had made their way to the Operations Building. After everyone had assembled, Colonel Coiner began the briefing.

"Men, you have all made it to Boringuen with no problems, and I am pleased with your performance, but we have some difficult segments ahead of us. Today's flight will take us to Trinidad, Port of Spain. We will fly an easterly course to San Juan then southeast to Trinidad. Be sure you take down all your check points and headings. The weather is clear with scattered cumulus, and possible rain at the lower altitudes. You have been given all the necessary radio and emergency frequencies, and we will maintain radio silence from here to our destination at Port of Spain. Also we have been asked to report any submarine sightings, as there are plenty of enemy subs operating in this area. Take off is 0700. You're on your own."

The trucks began to carry the officers out to the hanger area where their aircraft were waiting. The crew and crew chiefs were already there making preparations for the flight. The crew chiefs were checking over their ships, running the engines, and making sure they were ready for the long flight.

Captain West called out to his crew chief, "Robbie, how did she check out?"

"Fine, Sir. She's okay."

"All right, load 'em up and let's go."

"Bill, you take a check on the instrument panel, and I'll move her out."

Soon the engines were turning at 1,000 RPM, and Quinn moved the throttles forward to taxi out to the main runway. They were lining up quickly and angling towards the runway for engine checks, then as quick as one plane took off another took its place. At 30 second intervals for take off, the whole process would take nearly 30 minutes. Quinn was giving his engines a final run.

"Okay, she sounds good. Lt. Silverbach's plane has just pulled onto the runway. We'll follow him."

West's ship took off, and turned in a slow arc toward the east. Puerto Rico was a beautiful island with lush green sugar cane plantations and small farms dotting the land everywhere. It was especially beautiful this morning, as the early rays of the morning sun glinted on the silver wings of his aircraft. The scenery was beautiful with mountains and green forests covering the central part of the land. It was a short while before the city of San Juan came into view, with its orderly streets and crowded city dwellings.

As the aircraft turned southward toward Trinidad, Quinn began to gain altitude and leveled off at 8,000 feet. At that height everyone enjoyed seeing many of the coastal islands of Puerto Rico.

"Well, Bill, this is the last view."

"Sir?"

"This is the last American soil we'll touch for a long time."

"Yes, Sir, I hope we can get this war over with soon and get back home again."

"Yeah, I'm sure going to miss my wife and son while I'm gone."

"How old is your boy now, Capt'n?"

"Nearly a year old, and he just started walking a few months ago. He knows his daddy, too. He always gives me a big grin when he sees me come home now. You don't know how good that makes me feel to see that little kid smile at me. There will be so much I'll miss by not being there to watch him grow up, but I guess I have a real consolation in knowing that if anything happens to me, he will carry on my family name for me."

"Aw, Capt'n, don't even think that way. We're all going to make it through, every last one of us. You remember Ruby telling us at the banquet she knew we would all come back."

"Ruby is quite a gal. She's never left my side since we got married three years ago in Ashville. All through basic training, flight school, MacDill, and Hunter Field, she has followed me, and lived in some mighty bad housing arrangements occasionally, but she never complained. Lots of times I've had to stay on the base, and leave her for weeks at a time, and all she's ever said was, 'Honey, it's worth it all if we just get a few days together every now and then.' I've never doubted for a minute that God sent her to me because she's been such a perfect wife for me and such a wonderful mother for Johnny."

The plane droned on with the now familiar vibrating roar of the engines, and soon Quinn had motioned for his copilot to take over the controls for a while. Then after several hours Trinidad was nearing.

"Lt. Hunsicker, what's our ground speed?"

"About 210 MPH, Captain, we've got a little tailwind."

"Good, is our ETA still 1200 hours?"

"Yes, Sir. We should be passing the Windward Islands soon, and then Trinidad."

The plane was performing perfectly and the crew was settling down to a routine that weathered the boredom of the long flights.

"Lt., what's our radio compass look like?"

"On course, Sir."

"Bill, isn't that the island just ahead?"

"Yes, we're on target again."

"Okay, give the tower a call and I'll turn her into the landing pattern."

It was just a matter of minutes and the aircraft was taxiing to the apron and parked with the others.

"Bill, you and the men get checked into quarters and I'll check in at Operations and meet you in the mess hall."

Everything had gone fine so far. The men had gotten a good meal that evening and had gone to their barracks for the night.

The next morning after breakfast they all had gathered in the Operation's Office for a briefing from Colonel Coiner.

"Men, this will be a short leg of our trip. We will skirt the coast of Venezuela and land at Georgetown, British Guiana. Keep a careful watch for check points, as our weather may be rough the nearer we get to Atkinson Field. Take off at 0700."

The take off was routine and the crew was settling down to a pattern. Hunsicker's calculations were all proving to be accurate, and this was a relief to all, as they talked and tried to stay busy to weather the boredom of the long flights each day. Robbie and Chester would sometimes try to look for other aircraft in the Group and occasionally they would spot a few but the cloud cover and flying at random altitudes seemed to keep them from seeing very many.

It seemed like they had flown for hours, just seeing the dark green jungles of Venezuela below on one side and the ocean on the other side. Quinn had already made up his mind that if he had to ditch, it would be in the ocean near the shore and not in the dense jungle. It looked like you could wander for months down there and never find a village.

"Lt., what's our ETA on present heading?"

"We've just passed our last check point, and Atkinson Field is dead ahead, Capt'n, about 30 minutes flight time."

"Yeah, it won't be long before we will be seeing the outskirts of Georgetown."

It didn't seem very long before Captain West's intercom came on."Bill, start letting down to 1,000 feet and I'll call the tower for landing instruction ... Atkinson Tower, this is aircraft number 138 on beacon frequency at 1,000 feet, three miles northwest of field."

"Number 138, proceed on course, turn west on final to runway two. Over."

"Number 138 to tower, Wilco. Over."

In a few minutes they had touched down on the runway and taxied to the hanger area. They could quickly see that the accommodations here were going to be a little rougher than usual. The hangers and the out shops looked a little weather beaten and the barracks were quonset huts for quarters. No one seemed to mind; however, as all they wanted was to check into Operations, get assigned to quarters, find a good meal in the mess hall, and relax in their barracks before bedtime.There were some extra nuisances since they hit the tropics. Mosquito netting at night, and atabrine tablets to ward off attacks of malaria, as well as laying in bed and listening to eerie sounds of the jungle at night. They were acutely reminded they were a long way from home when they heard many of the strange noises in the jungle area surrounding the far side of the Field.

The procedures during the following mornings after each flight were getting to be routine. That was, breakfast at 0530, briefing at 0600, and take off at 0700. This time they were late for the briefing because of a late breakfast, but it wouldn't delay take off time.

As usual Col. Coiner was giving the briefing details.

"Take off time at 0700, men, and this time we will have plenty of check points by skirting the coasts of Dutch and French Guiana. This is about a four hour leg of 812 miles to Belem, Brazil. We will run into some bad weather this trip. Rain and solid overcast out of Atkinson. You may be on instruments a good part of the way, and you may have difficulty picking up your check points visibly. Start your let down about an hour before ETA at Belem, and you should be able to get a visual at about 1,000 feet over the Amazon Delta. Navigators, you will have some check points over Marajo Island, and then pick up Belem southeast of the delta area. Maintain radio silence and keep an eye on the coastal waters for subs. Any questions? ... Dismissed."

The men went to their aircraft with a feeling of uneasiness. The weather didn't look good at all, and now the navigators would have to sweat for visual check points. It looked tough.

Quinn started his engines and began to taxi out in line with the other planes of his squadron. One by one, minute by minute, the sky filled with the thunder of aircraft engines with planes grabbing for altitude and swinging on course to Belem, Brazil.

"Captain West to crew, keep alert on this trip, men, we're going into some rough weather in a few hours."

The weather looked fine at take off and the sunrise had been beautiful and clear, but as the flight winged its way to French Guiana the weather began to slowly close in. Quinn climbed to 9,000 to see if he could break clear, but it was no use. The plane was getting a little difficult to hold on course with so many up and down drafts buffeting it from different directions.

"Hunsicker, we're on instruments and can't see a thing from up here. Have you got a fix on the beacon at Cayenne?"

"Yes, Sir, we're on course. Just hold the same heading, and we'll probably break through at let down before we get to Belem."

The flight seemed to last forever, but Hunsicker had guessed correctly, for as they began to let down it began to clear at 1,000. "There she is, men, look at the Amazon River, the longest river in the world and probably the widest at this point in the delta. We should be coming into Belem in about 30 minutes or so."

"Lt., what's our ETA for the airfield?"

"1235 hours, Sir, we had some bad headwinds back over French Guiana, but we're on course and I'm hearing Belem's signal loud and clear."

"Bill, this has been a long flight for me. I'm worn out."

"I think we all are, Capt'n. This has been a rough segment."

"Well, there it is, the outskirts of Belem. Give the tower a call, and I'll take her on in."

After taxiing to the main hangers, Quinn parked the ship and headed to the Operations Shack. It wasn't a very large place, and it seemed that the further they got from home, the more primitive the air fields became. Now and then, they would see a group of native workers around the air base. It was always interesting to see their manner of dress and hear them speaking in different languages. Sometimes a little group of adventurous children would wander on the base and beg for candy. They were more interesting than the older natives, and most of them had learned some basic words to get goodies from the American airmen.

Quinn's crew left the operations area and were heading for the mess hall when they saw a small bunch of little kids calling to them.

"So'jer men, plezz, candee, plezz."

Quinn smiled and walked over to them as he turned to the others in his crew.

"Okay, Natanek, see if you can dig some candy out of my B-4 bag, and we'll give these kids a treat."

"Kids, can you speak English?"

"Noooh Englezz. Candee pleezz."

"Okay, so you can't speak English and I can't speak Portuguese."

Quinn tried to find out their age by asking and holding up four fingers and three fingers, but they just didn't understand.

"Who is the little timid girl back behind you?" Quinn asked, pointing to the little child.

"She seester."

"Oh, I see. She's your sister."

Quinn motioned for the little girl to come closer. She was a poor, little street urchin with torn clothes and a dirty face, but Quinn's heart reached out to her.

"Come here, darling, this candy is for you and your brothers. Now, skedaddle before you kids get thrown off the base."

The kids ran like little rabbits across the hanger area. Natanek said, "Did you see the smile on that little kid's face when you gave her the candy?"

"Yeah, it was beautiful, wasn't it? Kids are just something special, no matter where they come from. They all seem to have a happy heart when they know you care for them."

"Capt'n, look at those palm trees. This is really something."

"It sure does look like a page out of National Geographic, doesn't it? We'll be seeing plenty of scenery like that for a while, because we are right in the middle of the tropics now. They're probably still having some snow in the northern U.S. and here it's hot and humid like summer. I don't think the temperature changes very much here winter or summer."

"Are we going to have another early morning wake up?"

"Sure thing, probably about 0500 for chow, then 0600 for the briefing session. I know it's getting monotonous, but after this next segment the rumor is we are going to get a few days rest."

"Hey, that will be great! I'm getting worn out with just the same routine every day. Maybe we can take in a movie at our next base."

"Well, if they don't have a movie, they'll probably have an Enlisted Men's Club and you guys can listen to some music and relax. See you all tomorrow, and let's hope the weather looks good for us."

The men decided to look for the Enlisted Men's Club to see what activities were available. As they were walking, Robbie was speaking to Chester. "Chester, do you think our mail will ever catch up with us when we get to England?"

"Yeah, it shouldn't be long before we will be getting mail. I'll be ready for it too. A few letters from home should give us a boost." The Club was easy to find but it was small and not very interesting. After a few hours they walked to their barracks and got a much needed night's rest.

The briefing session the next morning was just that, brief... It looked as if everyone was ready for a change of routine, even Col. Coiner.

"Men, I'll make it short today. We'll skirt the Brazilian coast for 800 miles to Natal, Brazil. The weather is cooperating today. Visibility unlimited with widely scattered cumulus. Cruise above 6,000 feet and you will be in the clear. You have your check points and emergency frequencies. I know this is getting a little tiresome but we'll have a three day lay over scheduled in Natal to get ready for the long jump to Ascension. That's it, and good luck."

Dawn was just breaking as the aircraft lined the runway for take off. The morning sky was tinged a beautiful orange-red color, and the first rays of the morning sun were beginning to spread early light over the green terrain. The sun had not yet broken over the horizon, but as the planes would take off and gain altitude, the most beautiful of panoramas would appear; a new sun and a new day.

Quinn pushed the throttles forward for take off and the plane rolled down the runway, lifting off slowly, then it began its ascent to 9,000 feet. The weather reports were accurate as usual. At 6,000, the scattered cumulus began to top out, and it was clear from that point on. They were cruising at 185 MPH indicated airspeed with an estimated flying time of five hours to Natal.

Two hours into the flight, Sgt. Natanek's voice came over the intercom, "Captain, there is a large city over to the right. That should be our check point of Fortaleza."

Hunsicker replied, "Yeah, that's her, right on course. We are getting a strong signal from the beacon in Natal, so it won't be much longer until we are sight seeing in town."

"Bill, how about taking over the controls for a while, and I'll take a break. Keep her on 102 degrees heading, and I'll lean back and take a nap."

"Yes, Sir." Bill added, "Captain, I'm beginning to miss the rest of the crew."

"Yeah, I am too. Those guys can find humor in everything they see. You know, their troopship could reach England nearly as soon as we will. While we take a three day lay over, that ship is going to cover nearly 2,000 miles."

"You're right. They are moving slow, but they are plowing ahead day and night. Captain, you remember how Zola gives nicknames to everyone? Well, I have to laugh every time I think about what he said one day. They were talking about you, and Zola said that you could land on a dime and give nine cents change. That was funny, but he called you 'John Q.' when he told the story, and that's what made everyone grin when they heard it. How do you like that for a nickname?"

"Hmmm, well, it could have been worse. Anyway, he complimented my flying ability ... That's a plus ... You know something? We really need short names for the intercom when the fast action begins. Sergeant Picklesimer is going to get shortened to 'Pic,' and Natanek is going to be nicknamed 'Nat.' I guess everyone else is okay except Lt. Hunsicker. I haven't thought of a name for him."

Captain West clicked on the intercom and spoke, "Lt. Hunsicker, what is our ETA to Natal?"

"We're about 55 minutes out from the airfield, Capt'n, with an ETA of 1335."

"All right, that sounds good to me. I could use a little rest. How about you guys?"

They all agreed with West. Nearly 4,000 miles in a Marauder would earn anyone a rest from the constant drone of engines and the monotony of ocean flying.

The engines were still performing flawlessly. They had run perfectly the whole trip, and every move of the mixture controls and throttles had given a good response from the pair of Pratt and Whitneys.

Quinn soon eased back on the throttle controls and began the slow let down to the airfield at Natal.

"Bill, give the tower a call and get some landing instructions."

"Yes, Sir, there she is dead ahead."

A slow turn to the right put them in the landing pattern and in a matter of minutes they were on the ground and taxiing to the group of hangers at the end of the runway.

"Okay, men, pile out. Col. Coiner's plane is on the apron. Steere and Allen made it in and that looks like North and Bergman coming in now."

"Capt'n, do you think you can get a jeep from the motor pool and give us a ride into Natal?"

"We'll have to wait and see if we are restricted to the base or not."

"Man, it would be great if we could see part of the town before we have to take off again."

"I doubt if we can get anything at the motor pool. It will be more like 'catch an army truck into town' if we can get passes to go off base."

Quinn checked into the Operations Building, filed his flight reports, and asked the desk clerk, "Sergeant, what are the regulations on going into town?"

"It's all right, Sir. Your group isn't restricted to the air base."

"What about transportation?"

"There is a truck that leaves the airfield every hour, or you can get a ride on any of the G.I vehicles heading to Natal."

"Okay, Sergeant, we thank you."

Quinn and his crew checked into their quarters and headed for the mess hall. Later they caught a ride on a truck and made their way to Natal. It was an unusual city with palm trees lining some of the main streets and little shops everywhere. The most interesting sights were the people. Bronzed natives with strange dialects gathered in groups to barter and chatter. It was a strange sight for the young Americans.

The little street shops were interesting and numerous. Most were just covered with tent material and vendors spread their goods under the shade to hawk their wares. Others were completely open and displayed farm products, fruits, and craft items of brass, copper, and silver. Some of the more boisterous merchants would call to the men and hold up items of trade to tempt the men to buy souvenirs.

"Hey, Joe, one dollar. You buy?"

Natanek wanted one of the items for his girl friend back in the States.

"Captain, do you think it's a good buy for a dollar?"

"Yeah, it looks nice to me. Just remember what they told us at the base. The natives only know one price, and that is 'one dollar' so whether it is cheap or expensive the price is still a 'buck.' Look carefully and pick out the nicer stuff."

While they were making up their minds about buying, Quinn felt a tug on his coat sleeve. He looked around and saw a tiny native boy with a broad smile looking up at him.

"I guide you, Joe. Me, Carlos, best guide in Natal."

Quinn looked down at the boy and smiled, "How old are you, Carlos?"

"Eight, Sir."

"How did you learn English so well?"

"My mother works at air base. She teach me."

"Okay, Carlos, you can be my friend. Where are some shops that have gold rings and women's jewelry?"

"Joolree? Oh, I see now, women's trinkets. You come. I show you. Is your name Joe?"

"No, Carlos. It's Captain West."

"Captain Westa, you my big friend. Come all of you. I show you."

Several streets distant Carlos showed them a large shop in a building area.

"You go in. I wait for you here."

It was a beautiful shop with every kind of jewelry and giftware imaginable. The shopkeeper spoke English and the prices were reasonable compared to the American shops, but these were not the dollar give aways like the smaller vendors were selling. The men had no trouble buying all the souvenirs they wanted and when they came out Carlos asked, "You buy nice things?"

"Yes, that was really a fine shop."

"See? Carlos knows everywhere in Natal. I take you to docks and we see ships load bananas and coffee beans, or we go to movie show, or Carlos show you good eat place. Best fish and rice in town, okay?"

Carlos was an untiring guide and he had nearly worn the men out from touring the town. Quinn recognized the street where they had gotten off the army truck and he hoped they might find a ride back to the air field.

"Carlos, can we catch a ride back to the field from here?"

"Sure, Captain Westa. Always soldier boys wait here and trucks stop to take them home."

"Carlos, I want to give you something. Is a dollar okay?"

"Sure, okay, plenty okay. You come back tomorrow. Carlos look for you. G.I. Joe truck come soon. You wait."

The men stayed busy for the remaining days checking equipment and going over instructions for the long flight to Ascension. Quinn had gone to the Base Post Exchange several times and on their last day in Natal he had picked up a gift for Carlos. The problem now was to try and find a way to get the present to him. He decided to ask some of the workers in the PX if they knew the little boy and to his surprise some of them knew him.

"Carlos? Sure, we know him. His mother works here at the PX. His name is Carlos Santos and the family lives in town."

Quinn asked, "Would you mind doing me a big favor? My Group is flying out tomorrow and I can't get back into town. Would you give this present to him?"

"Sure thing. We'll take it to him tomorrow."

The present was not wrapped so the men could easily see that it was a small book. They waited until Captain West had walked away and curiosity began to get the best of them. They slowly opened the book to the title page, and in Portuguese were the words "New Testament." The inscription was hand written, "To Carlos, my little brother, from Captain West, 1944."

On the morning of the Group's departure the briefing for the long flight to Ascension Island was at 0600, the usual early hour, but there were many facts to cover. As all of them assembled, Col. Coiner began to speak, "Men, Ascension Island is 1,200 nautical miles from here, just a speck in the ocean, only 34 square miles in area. You will be flying over nothing but ocean with no check points and no alternate fields for eight hours. You either make it to Ascension or ditch in the ocean ... We are going to give you some very detailed instructions about winds aloft at several altitudes, beacon and emergency frequencies, and some peculiarities of this particular segment. A word of warning about beacon frequencies. There have been times in the past that a German submarine has been hanging several miles off the coast of Ascension and beaming out the beacon frequency, so be sure you back your radio compass with some good dead reckoning. You have about nine hours of fuel for an eight hour trip, so you may want to fine tune your engines after take off by easing back the RPM's and pulling the mixture controls back a little."

You men have done a good job so far, and this leg is the proving ground for all you have learned in the flight schools. Navigators, check and recheck your course readings, estimated time of arrival, and beacon headings. Now for the pilots, you have two things to be cautious about when you get into the landing pattern at Ascension. There are thousands of birds that nest on the Island. The people there call them "Gooney Birds," and they can cause lots of trouble if you have to fly through them. Also be careful of the runway. It was cut out of solid lava rock and it is higher in the middle than on either end.

When you land be sure you are down solid in the first half, because if you are not, you will think you are airborne again when the runway starts dropping off ... Accommodations are poor as might be expected. There is little water supply, meager food supplies, and poor sleeping conditions. All we want to do is land, gas up, sleep, and get out early the next day after your breakfast of powdered eggs and powdered milk ... Good luck, men. Take off is at 0700."

The men were excited. It was a challenge to be able to hit that island dead on target. During the weeks at Morrison Field the airmen had made a song out of the difficulties of hitting the island. They would kid each other by singing, "If we don't make it to Ascension, our wives will get a pension." It was something to think about now that the time was at hand.

The dawn was gray as the planes took off one by one with determined men set on a goal of eight hours flying time to their destination at Ascension. At 6,000 feet they began to break out of the cloud cover, and the pilots began to fly with a little less tension. To fly in a group of 60 planes through thick cloud cover meant that there was always a chance of midair collisions, one of the worst of aircraft accidents. After several hours out of Natal the clouds began to dissipate, then scatter, and navigators breathed easier and began to show some signs of optimism. Now, wind and drift readings were possible to recheck the course headings and ETA's. The beacon signals began coming in stronger from Ascension. That was a comforting thought, unless they might be coming from that German submarine lying several miles off the coast, luring planes away from Ascension, and causing them to ditch in the ocean.

The navigators had another method of rechecking their course; by celestial navigation. When they weren't taking drift readings, they could periodically take some "sun shots" with a sextant, and double check course headings and ETA so they would know the exact time to look for the appearance of the island.

The flight droned on for hours. Long monotonous hours of nothing to look at except a bleak flat ocean. Occasionally the monotony would be broken by a crew member's interjection over the intercom and that was always a welcome change. Captain West's voice came over the intercom, "Hunsicker, what's our ETA now?"

"1408, Captain. Drift readings are showing an eighteen mile an hour head wind. Indicated airspeed of 228 MPH and ground speed of 210 MPH. We are supposed to pick up a tail wind during the last few hundred miles before reaching Ascension. I hope the meteorologist is right. We could really use some help to get us there a little sooner."

"Nat, are we still getting a strong signal from the beacon on the island?"
"Yes, Sir, loud and clear."
"Bill, you take the wheel for awhile. I'm going to lean back and rest."
"Yes, Sir. Captain? What's our next stop after Ascension?"
"Probably Liberia, but that could change. It will be somewhere on the African coast, then across the Sahara Desert to French Morocco, and on to England. About 10,000 miles. Isn't that something?"
"You bet it is."

The flight was almost hypnotic with the constant roar of engine noise and the absence of landscape to break the dismal sameness of the vast Atlantic Ocean. Hours followed hours until a familiar click of the intercom, and Hunsicker's voice came in.

"Captain, we are showing an ETA of 1400, so we should be seeing the island in about 30 minutes."

"Okay, Hunsicker, keep a look out for a small land mass. I'm going to start a slow let down now. Nat, you and Robbie get to one of the windows and yell when you see land."

The men didn't have to be told to look as each one was looking eagerly for the small island. Every cloud shadow on the ocean seemed to be the landfall they were looking for until getting closer they discovered it was only a shadow. Time and time again they were disappointed in thinking they had seen land until at last Quinn's copilot exclaimed, "Look to the right at one o'clock, Captain, is that it?"

"Sure looks like her, Bill. Give the tower a call and let's try that bowed-up runway."

Ascension Tower came in clear, "Number 138 clear to land. Come in on north end of runway. Wind is 15 MPH south-southeast."

"Okay, Bill, let's take her in. Gear down ... Full flaps ... Carb heat on ... Cowl flaps open ... Turning into final now ... Steady ..." screech ... screech. That familiar squeal of tires touching the runway told the crew they had made it to Ascension. Robbie's voice resounded from the midsection of the Marauder, "Hey, we made it."

"Yeah, it's a good feeling isn't it?"

They could see there were quite a few of the 397th Group planes already parked on a large apron off the center of the runway. Captain West was turning into the taxiway and into the parking area. As they began to pick up their B-4 bags and personal gear, Captain West reminded his crew, "All right, men, there won't be much trouble you can get into here, but check in and find out what time they have scheduled tomorrow's briefing before you hit the sack."

The men were curious about the field as they passed a large sign at Operations which read, "AIR TRANSPORT COMMAND, WIDEAWAKE FIELD ... ASCENSION ISLAND ... ELEVATION 278.29 FEET ABOVE SEA LEVEL ..." They later discovered that the birds which nested near the end of the runway were called "Wideawakes" and also that they were only a problem during the months of August and September. The island looked so curious with volcanic rock everywhere and large mountains of lava rock looming in the distant edges of the Island. It gave one the feeling that this place had just blown up from the ocean's floor and could just as easily sink back down again.

The oceanographers and geologists had said that this was the youngest of all the Atlantic islands and certainly had been formed by an underwater volcanic explosion which spewed the island up at a comparatively young date. It might give the transients an uncomfortable feeling, but any dry place in the middle of the Atlantic was a welcome resting place after 1,448 air miles from Brazil.

The pyramidal tents and army cots were not the best in the world but they had clean sheets. The men were tired so there weren't many complaints.

Wake up call was early the next morning at 0500, chow at 0530, and briefing at 0600. Colonel Coiner was smiling as he began the briefing session.

"We all made it to Ascension. No one had to 'ditch' in the ocean. That's a credit to you men. Now we have another leg nearly as bad as the one getting here. It's 886 miles to the coast of Africa, Roberts Field, Liberia. There will be more of the same conditions we had while flying to Ascension. All ocean, no check points, no alternate fields, with the exception that we will be aiming at a larger land mass. The weather conditions are not good. There is haze all the way to Roberts Field. You navigators will have to make sure your dead reckoning is near perfect to get a check on the beacon signals. Pilots be cautious on take off past midfield as the runway begins to drop, you'll get a feeling of being airborne, so don't pop your wheels up too quick. Just get a little altitude and be sure you are climbing before you retract the gears. Our take off time is 0700. See you men at Roberts Field."

The weather really didn't look good at all and with such limited visibility the navigators wouldn't be able to take drift readings to confirm wind speed and direction. It was just another chance for errors to creep in and navigational errors could be deadly at this point. The men would be glad to get Ascension behind them but it was still serious business getting to the African coast.

The aircraft took off at 60 second intervals and each one climbed to the prearranged altitude of near 9,000 feet. The aircraft were supposed to be flying close to 9,000 feet as suggested in the briefing but Quinn's plane had not broken out of the overcast after several hours flying time so the men just settled back and flew on instruments for the next few hours. Robbie came to the cockpit area to talk to Capt'n West and Lt. Budge.

"You know Captain, I thought we would see more of the other aircraft in our group but we have seen very few."

"Well, that's mainly because of our 60 second take off intervals. One minute puts the plane in front of us three miles away so we are all strung out at different altitudes and three miles apart."

"Okay, I understand now."

Captain West clicked on his intercom, "Hunsicker, are we getting a beacon signal from Roberts?"

"Yes, Sir, she's coming in clear and getting stronger all the time. The radio compass is swinging dead ahead so we should be on course."

"What's our ETA?"

"Captain, as far as I can calculate we should be over the field at 1136."

"Okay, that sounds good ... Bill, take over for awhile, I'm going to take a break."

The flight droned on with no let up in the weather and a kind of uneasiness began to settle in with so many unknowns in the navigational area. West had taken over the controls again and began a slow let down to see if they could break out into clear weather before sighting Roberts Field. He called over the mike.

"Hunsicker, give us some good news. We're down to 2,000 feet and nothing sighted yet."

"Sir, we're getting a strong signal from Roberts right on course and our ETA is still 30 minutes away."

West kept his eyes on the clouds ahead looking for a clearing as he talked to his copilot, "Bill, give the tower a call for weather and landing instructions."

"Capt'n, isn't that Roberts ahead?"

"Yes, there she is. It sure helps to get a little break in the weather. We'll need to make several turns to get into the pattern. Look at those villages below with nothing but mud huts. If we thought Ascension was bad, it might have been a picnic compared to this place."

In a few minutes Quinn was wheeling his aircraft to the apron where several other Marauders were parked."Okay, men, all out. I hope this place has some good chow. I'm starved."

"Me too, Capt'n," Robbie replied.

After the men had eaten and found their quarters it was sack time, a little rest, some letter writing, evening chow, bunk down, and get some sleep until wake-up call at 0600 the next morning.

The briefings were getting tiresome but that was the way it had to be. The long flights were a constant drain on their nerves, but no one complained. It was just accepted as a vital part of the war effort. It was important and vital for victory. They were young men anxious to win a war so they took it all as a part of their job.

Each segment they flew was drawing them closer to their combat stations and this was bringing a mixture of emotions. They were outwardly eager to get into the action overseas but inwardly was the nagging uncertainty of battle. They knew they were taking a gamble with their lives at stake but this is what they had been trained for. This was their mission and they were confident that someway they would pull through it all.

Colonel Coiner was a man who could give an airman a feeling of confidence. His toughness and brash exterior gave the men a will to emulate his tenacity. If he thought the men could do the impossible they would try double hard to prove that he was right.

That morning the briefing was going at a swift pace with Coiner giving the Group their headings and beacon frequencies. "Men, this leg will skirt the coast of Africa for 700 miles to Dakar, Senegal. You are not to go inland as there are some neutral countries we don't want to fly over. Haze may be a problem for several hours until you break out of it above 6,000 feet."

The flight was nearly routine with plenty of check points along the way. After the uncertainties of the long ocean flights this was a milk-run and the men were in high spirits.

Captain West's voice came over the intercom, "Hunsicker, what do you know about Dakar?"

"Not much, Sir, I was not a real student of geography. I wish now that I had been."

"Well, it won't be long before we get there. What's our ETA?"

"About 1045, Sir. We are about 40 minutes out of Dakar."

The flight was uneventful and it was one of the easiest legs of the journey. Only three hours flying time made it a joy to fly this segment and each step closer to England seemed to heighten the excitement of the Group.

The next day's briefing was more thorough than the few before. Col. Coiner was going in depth into some of the details.

"Men, you know the problems of over-flying neutral countries so our course will skirt around Spanish Sahara. The main problem of this leg is the Atlas Mountains. After take off we fly over Mauritania, then over the Sahara Desert, then into French Morocco, and through this pass in the Atlas Mountains to our destination at Marrakech. These mountains are large. The highest peak is 13,661 feet so you must have good visibility to go through the pass. If it is cloudy, don't try the pass but turn west to the end of the mountain range by picking up these new headings here on the blackboard and double back to Marrakech. There is an alternate field at Casablanca. You will be given all beacon and emergency frequencies as well as all course headings and weather aloft at various altitudes. Take off at 0800."

It sounded exciting going places the men had never seen but had heard the names. They had seen movies about these strange and distant places — the Sahara, Casablanca, and Tindouf, which was an old French Foreign Legion Post in Algeria.

This leg would be over land and the men were anxious to see the great

Sahara Desert. After a few hours out from Dakar, Captain West announced, "Men, take a good look below. We're over the Sahara."

"Captain, it looks pretty flat to me."

"That's just the way it looks from 9,000 feet but it must be rough down there. I'd hate to try and land on it."

"Hunsicker, what's our ETA and present position?"

"Capt'n, we're 100 miles from our next check point at Tindouf, Algeria. There's an emergency field down there and it's our last check point before getting on course for the Atlas Mountain pass. Our ETA for Marrakech is 1245."

Robbie answered, "Hey, maybe the pass will be clouded and we can go around the mountains and land at Casablanca. Do you remember that movie?"

"I sure do but the town is probably not much like the movie version. Anyway Marrakech will be just as intriguing as Casablanca."

The long flight continued and at last the Atlas Mountains began to appear on the horizon.

"Bill, take a look at those mountains. Aren't they beautiful? They look nearly purple in the distance and with those snow capped peaks, that's really a sight to behold."

"Yes, it sure looks like a picture postcard in color. I wish we had a camera to catch that shot but no camera could do it justice."

"Men, there's the Atlas Mountains straight ahead and it looks like the pass will be clear. We'll be picking up some more altitude as we get closer. Hunsicker, what's our course?"

"Sir, we're on course, and if we hold our present heading it will take us directly through the pass."

"Robbie, come up here a minute. Look far ahead and slightly off to the left. Isn't that one of our ships?"

"Yes, Sir. I see it now. That's one of our Marauders."

"Okay, you men may want to hook up to oxygen. We're going to push on to at least 12,000 feet. Those peaks look more rugged by the minute ... There's the pass ... We should be going through in a few minutes."

"Nat, take a look at those sheer bluffs of granite. It really seems strange flying through this pass and seeing mountains on each side of you. Are you getting a beacon signal from Marrakech?"

"Yes, Sir, it should get louder when we break out of these mountains."

"I'm starting a slow let down, Bill, and turning left 18 degrees for our new heading. Nat, how is the radio compass looking?"

"Looking good, Captain, she's reading dead ahead with a beacon frequency loud and clear."

"Bill, we should be in range of Marrakech Tower. Go ahead and give them a call."

"Marrakech Tower, this is number 138, in flight group 397, ten miles north, northwest. Request landing instructions. Over."

"Number 138 proceed on course to east of runway one. Winds 15 miles per hour westerly. Check in again at taxiway. Over."

"Number 138 ... Roger ... Wilco ... And out."

"There she is, Bill, a left turn will put us in the pattern, and two turns will put us on final."

It was all over in a few minutes and the men were piling out of the plane and wondering what the town of Marrakech would look like.

Robbie was exuberant, "Hey, Nat, the first thing we need to do is get some chow and then head to some of those street markets in the 'Casbah.' You can probably pick up some jewelry for your girl friend down there real cheap."

"Maybe so, but they say to stay clear of some of the older sections. They are pretty dangerous. When we check into Operations they can probably give us the low down on it."

The Operations Building was buzzing with activity as everyone had questions about passes into town. The Desk Sergeant stood up, "Okay, you guys hold it for a minute and I'll give you the run down on Marrakech. You are not restricted and you can all get passes into the main part of town but don't go alone. Go in a group. It can be dangerous in some of those side streets. Some of these people have little respect for that uniform you are wearing and less respect for the country which you came from. Also, the old town of Medina is off limits. Servicemen have disappeared out there without a trace, so be careful."

Robbie laughed as they were leaving the building. "Captain, did you hear that? That's where we need to go to — Medina."

"No, you guys are going to stick with me. I've taken you this far and I'm not about to lose you on some trash heap in Medina. We'll get plenty of sightseeing done in Marrakech without taking a lot of crazy chances."

The trip into town was exciting with hundreds of small bazaars and people of many cultures. There were those of Arab descent, Moors, Berbers, and Europeans. Most all were Muslin and the hundreds of Mosques in the town dictated this faith. Men with turbans, merchants with red fez and Moorish costume, but also the women with long skirted costume, and covered faces veiled except for the eyes, were a sight that the men had seldom seen except in school geography books. It was like another world with the Arabic language of the merchants and shop owners chattering over a business deal.

Natanek was walking beside West and Hunsicker and taking in all the sights.

"Captain, will we be here for a few days or are we scheduled to leave tomorrow?"

"We may be here a few days. I heard Col. Coiner talking to Major Dempster about winds aloft being too heavy for us to make the long jump to England tomorrow. Don't spend all your money too quickly. You'll more than likely be able to come back the next few days and bargain with some of these merchants."

Quinn's guess had been right but the days passed swiftly and soon it was time for the briefing. The men were again anxious to get in the air and head for England. Col. Coiner's voice boomed from the front of the gathering.

"Men, this is the last leg of our journey and one of the longest. It could be one of the most dangerous for several reasons. First, it is a 1,340 mile leg to Lands End, England. Secondly, you will be flying near enough to occupied France for German fighter aircraft to intercept your Group. So be extremely cautious and watchful. Load your guns, maintain radio silence, and turn off all lights after leaving Marrakech. This time we will fly rather loose squadron formations for protection against German fighters coming out of the Brest Peninsula of France. We are skirting around the coasts of Portugal and Spain because these are neutral countries. If you have to make a forced landing, destroy all secret documents. Give only name, rank, and serial number and ask to see the American consulate. When we near England our landing approach to Lands End is only allowed from the north. The code name for the airport there is 'Odd Job.' You will have frequencies for that station and emergency frequencies for rescue if you have to ditch before reaching England. Men, take off is 0700 and you will be entering the combat zone. So, good luck to you all."

It was a sobering thought as they took off to realize that in a few hours they would be in range of the German Luftwaffe.

There was an elite German Fighter Group operating out of Abbeyville, France, and they were a force to be avoided at this point in time. The flights droned past Portugal, then Spain, and the course pointed north toward Ireland, far out from the coasts of France. It seemed to be a round about way of getting there and the pilots and crew chiefs began to worry about fuel consumption.

Captain West called over the mike, "Robbie, come up here a minute. I'm going to ease the mixture control back a little to see if we can't squeeze our fuel supply. Listen for a cough in the engines and we'll ease it forward enough to keep them running smooth. Maybe we can hold our fuel consumption to a minimum."

Hunsicker clicked on the intercom, "Captain, turn south in ten minutes to our new course and we'll be heading to 'Odd Job.' ETA is 1430 on present heading. We should be over Lands End in thirty minutes."

"Thanks, Hunsicker. That's the best news I've heard in weeks. Nat, are we homing in on Lands End?"

"Dead ahead, Sir, the signal is loud and clear."

"Okay men, just a little longer, and we'll be in England. Bill, try to raise the tower on the radio."

"Odd Job, this is number 138 in Group 397, ten miles north of station ... Request landing instructions ... Over."

"Number 138, continue on course to north end of runway ... Turn right onto final ... Winds south at 15 knots ... Over."

The men had been looking for land for the last 30 minutes and each one wanted to get their first glimpse of England.

Robbie shouted, "There she is Captain, off to the right. It looks beautiful, doesn't it?"

"Yes, it sure does. That's going to be our home for a while so take a good look at her."

"Capt'n, look at that rugged coast line. Just like high cliffs coming out of the ocean and that beautiful patchwork of green meadow. It really is something to see."

They didn't have much time to look, for Quinn had wheeled in on final and touched down before they realized it. They couldn't wait to check into the Operations Building and talk to some of the personnel there. Several officers and enlisted men were manning the station in R.A.F. uniform. The clerk-typist who waited on them was a corporal in the R.A.F. and also an attractive young lady. The men were all ears as Quinn asked, "Corporal, are there any instructions for the 397th Bomb Group?"

"Yes, Sir, there is a meeting scheduled at 1900 in the quonset building at the left of Operations ... Welcome to England, Sir."

"Thank you ... We're glad to be here." Quinn turned to leave and the men were still standing enthralled by the crisp English accent of the young lady. Quinn motioned for the men to follow him and smiled as he said, "Come on, men, we've got some work to do."

The meeting that night was a celebration but it was also an introduction of the things ahead as Col. Coiner came before the Group. "I don't have to tell you men how proud I am of you. Flying sixty planes for 10,000 miles without a major mishap. I'll give a toast to the best Air Force Group in the world ... The 397th."

After the shouting subsided he began his remarks, "We have some work ahead of us learning the 9th Air Force operational patterns. It will be a little different than we have been used to in the States. Also, we must familiarize ourselves with the English operations, codes, and emergency frequencies, so this meeting tonight is a brief orientation. Tomorrow we fly to our base at Gosfield. We'll stay there until our flight crews and ground crews join us sometime around April 1st. Until that time we will be training, preparing for bombing missions, flying practice missions, and getting our skills honed to a fine edge. Take off time is 0800 tomorrow. Have a good time tonight and celebrate our arrival in England."

The men were ready to celebrate after the long trip overseas. The hubbub of conversation was sometimes an uproar intermixed with laughter as every one was talking and telling about their experiences on the flight to England. The party continued until late that night when some officers began to talk about the work scheduled for the next day and the crowd began to break up and head to the barracks.

The work at Gosfield proceeded at a fast pace as there was much to be done. But this made the days fly and soon the date for the arrival of the ground crews was near.

Rumors were flying everywhere ... "They're coming in this evening ... No, they're coming in tomorrow ..." And so it went but the day finally arrived and the crews wheeled in to a roaring reception. Everyone had missed them and it looked like a great reunion with all the shouting, hand shaking, and back slapping.

Quinn and his crew could hardly wait to see Zola, Pickelsimer, and the other crews they had trained with. Robbie spotted them first and yelled, "Hey! Captain! There's Zola and Pickelsimer getting off the truck over there. Hey, Zola, man have we missed you guys. What took you so long?"

"Well, we just had a slow boat. We missed you guys too. How did the flight overseas come off, Captain West?"

"We did fine except we wished you and Picklesimer could have been with us when we flew over."

"Capt'n, I'll tell you the truth ... We wished a hundred times we could have been with you on that flight but we all made it here and that's all that counts."

"You said it. We'll all agree on that one ... But, we've got our crew together now and they'll have a hard time separating us again."

Robbie broke into the conversation, "Pic, you know we are having a party for you guys after chow tonight? Everybody is going to be there ... Be sure and tell Donzello and Skarles ... Tell Ray Snow to bring Kitrick and Sears with him ... Boy, we'll have a ball!"

Nat and Zola began to walk with the others over to the mess hall. Nat spoke first, "Zola, I sure am glad to see you again."

"Yeah, I didn't know how much this crew meant to me until we got separated."

After chow that night the men began to gather for the party. It was a party to remember. Everyone was there ... Wanous and Gauker ... Burns, Russell, and Rodie ... Spotkov and Thorp ... All the pilots and copilots; Stangle, North, Silverbach, Ryherd, and all the others. The bombardiers and navigators were there in force; Daoust, Lee, Cook, Bown, Casey, and all the gang. What a group! Everyone trying to talk at the same time made a hilarious uproar.

Col. Coiner and his staff officers were there and during a break in the noise of the festivities, Col. Coiner asked to speak, "Men, give me your attention for a minute. I want to personally say how glad I am that our unit is back to 100 percent again. We will be training here for a few more days but the good news is our orders have been cut for us to move to our permanent base at Station 168, Rivenhall, Essex, England, effective April 15.

The excitement and uproar began again as the men cheered their new base of operations near Chelmsford, England. Essex was ideal for the bombing missions of the Ninth Air Force. It was near the coast only a short hop across the Channel to German occupied France, where most of the targets were located so the Essex area was nearly covered with Ninth Air Force Airfields.

The announcement heightened the party atmosphere. The Group continued to renew acquaintances talking about old times at MacDill and Avon Park and the good times at Hunter Field and Morrison.

The days passed swiftly and soon the 397th had been indoctrinated into the flying procedures of the European Theatre of Operations and the Ninth Air Force. The baggage was moved by truck to Rivenhall and the air crews made the short hop to the airfield. It was a large place. The permanent quarters were quonset huts with prefabricated buildings predominating. There were several large aircraft hangers and maintenance shops. It was ideal for the Marauder Group. The reason for the completeness of the field was that there had been a Ninth Air Force Fighter Group operating out of it for several months. They were the 363rd Fighter Group flying P-51 Mustangs. The runways were more than a mile long which was needed for the medium bombers carrying a heavy bomb load. The taxiways and hardstand areas as well as the perimeter tracks were more than adequate.

For four days they unloaded baggage and set up operations on the field. On the fifth day they were scheduled for combat, April 20, 1944. The preparations were over and the proving ground began.

Standing from left, Capt. Steere, Lt. Lipscomb, and Sgt. Johnson. Kneeling are Sgt. McGinnis on left, Sgt. Schubin holding crew mascot, and Sgt. Mitchler.

CHAPTER SEVEN

COMBAT MISSIONS, LE PLOUY FERME TO OUISTREHAM

The men were all eager to begin the bombing operations since their long months of preparation had been getting them ready for this day. They had a feeling of confidence in their Group's ability and an inner assurance that everything was going to turn out for the best.

The first mission was set for April 20th, and always on the day before a mission a "Loading List" was posted on the board at Operations. The list had each pilot's name and the number of the aircraft which he would be flying. Also, it indicated the position of each aircraft in the flight pattern or flight boxes. Col. Coiner would lead the first box of eighteen aircraft and Major Dempster would lead the second box of eighteen. The armorers would be working most of the night to prepare these planes for combat. Bombs had to be loaded and fused, machine guns stripped, cleaned, mounted, and loaded with belts of ammo near each position. It was exacting work but there was no room for error when men's lives hung in the balance.

The crowd around the operations building was large as Zola and Natanek approached.

"Nat, do you think we are on the list for the first run?"

"I don't know, but I bet we are. Here comes Donzello ... Maybe he will know something."

"What's the verdict Donzello ... Are we on it?"

"Yep, you're on it buddy. Barnett is on it too, so we will be flying together in the first box. Coiner is leading and Col. Allen, Ryherd, Bronson, Quiggle, McLeod, Smith, and Taylor are all in our flight box."

"Any rumors about where we will be going?"

"Nope, all of that has been kept pretty quiet."

"Come on, Nat, let's push through and get a look at that list. I want to see West's name and old number 138-C on that thing."

The men were excited and sleep didn't come easy for them that night. There were so many variables to think about. The briefing was set for 0700 and it couldn't come soon enough for them as they were all anxious to know where they would be heading.

The next morning everyone was tense. The briefing room was large with a group of maps behind the speaker's platform and ample chairs for all the men. The noise of conversation and laughter seemed to rule the day until a voice in the rear of the room commanded, "Atten-hut," and all became quiet and stood at attention as Col. Coiner walked to the speaker's platform.

"At ease, men ... I know you are all anxious to know where we are going. Here it is on the map ... Le Plouy Ferme. The target is a V-1 rocket launching site northwest of town at coordinates 02022-34016. We are not sure at this time what these sites will be used for, but intelligence tells us that the Germans are getting ready to launch these rocket bombs on England and probably center on London and the surrounding area. Our job is to destroy them before they get started. The weather is not good over the target area with a 8/10 cumulus formation which is expected to slowly clear. Our Executive Officer has some more information for you. Take off time is 0800 ... I will lead the first box and Major Dempster will lead the second ... Any questions? Okay, good hunting."

The group executive officer gave some comments regarding coordinates for the IP, target, and return, then he introduced one of the S-2 intelligence officers.

"Men, this is Captain Taylor, and you will be seeing lots of him at the debriefing sessions following the missions, but for now he wants to give you some background on the type of target we will be flying against today." Captain Taylor took over.

"You may have been briefed before on these rocket bombs, but it is becoming more important for us to know the type of base and operation we are dealing with. We don't know much about these rockets except what the French resistance fighters have secured for us in the way of parts of the rocket and what our intelligence teams have discovered from aerial photos of the bomb sites themselves. We have code named these rockets as V-1's, and they seem to be Hitler's last ditch stand or at least his revenge bombs for the raining of destruction on England's major cities. They are designed to be fired from a long inclined ramp type of structure which is easily recognized from our aerial photos. Also in close proximity is a long building with a curved end which resembles a ski turned edgewise. Probably this building is used for storage of the bombs. Some use the term 'Ski Sites' for these locations because of their curious shape. All of these sites have some common features, that is, the ramp and the storage shed. What is particularly upsetting is that they all are pointed to London. There has been a great proliferation of these sites in the last few months and they must be neutralized. You men are a part of the operation 'Crossbow' which is our attempt to wipe out these locations before they are able to begin firing upon London."

Col. Coiner returned to the platform for some final words.

"Men, this is our first mission ... Let's make it a good one. You'll be loaded with eight 500 pound bombs ... Take off time is 0800 with a green flare signal from the control tower ... Fighter rendezvous at St. Catherine's Point at 0815 ... You have all of your emergency radio frequencies ... Radio silence after reaching the Channel ... Okay, get your gear and load up ... Good luck."

Nat and Zola filed out together and headed for their flight equipment, chute packs, and flight jackets. It was exciting but there was a gut feeling of facing the unknown. They were glad to be flying their first mission for they had trained these past months and years for this purpose, but there was always a nagging feeling of apprehension which follows all men into combat. Zola threw his gear into the truck with the others and squeezed in. It would be just a short ride to the hardstands where their squadron's aircraft were parked. The night armorer crews had loaded the 500 pound bombs into the bomb bays and loaded all the racks with ammo above each machine gun location. The crew chiefs had checked the planes over from stem to stern, checking engines, hydraulics, controls, and instruments. All the aircraft were in readiness.

The jeep pulled in with Quinn, Budge, and Lt. Hunsicker who would be their Bombardier-Navigator for this flight. West had gotten out when the truck pulled up with the crew. "Okay, Nat, Pic, climb in ... Zola, check her over good and give me a report before we take off ... Robbie, has everything checked out?"

"Yes, Sir, Capt'n, she's ready to go ... The engines sound good and all the hydraulics checked out."

"Okay, Budge, climb on up and I'll follow you."

Quinn took the left pilot seat and pulled back his window as he looked out at Robbie hooking up the external power supply.

"Robbie, is number 1 all clear?"

"Yes, Sir, she's clear."

Budge had already checked over the instruments, but West glanced at the engine instruments again. Mentally checking each one, master switch on, fuel valves on, generators on, propeller toggles to auto, throttles set for starting, prime left engine, energizer switch to left. The energizer began its whine as it picked up RPM's to a high pitched sound and West pushed the mesh switch to left position. The prop began slowly turning, then fired a few times and caught as it blew out a screen of white exhaust. Quinn held her back to 800 RPM's and signaled to Robbie. The same procedure was used on the right engine and in a few seconds both engines were running smoothly. He switched the fuel booster pumps to off, turned the battery switches on, and quickly signaled for Robbie to pull the external power source. A few more minutes and Quinn was ready to taxi out behind Col. Allen's ship. He eased the engines up to 1,000 RPM's and slowly pulled out of the hardstand area.

Captain West's voice came over the intercom, "We're three minutes until take off, men." The crew understood his meaning. It would become a tradition of the "By-Golly" crew, these few seconds before flight time, too recognize in silent prayer that a loving Almighty God directed their footsteps.

The planes began to taxi to the main runway and line up as they turned their engines to 2,400 RPM for a check of manifold pressure, magneto check, and propeller governor check. Then they turned onto the runway and waited for the signal flare. It was a matter of minutes as they saw the green flare and the planes began to take off at half minute intervals roaring down the runway. The noise of so many engines was deafening. Quinn watched Col. Allen move down the runway as he waited for a few extra seconds to be sure he would not be in Allen's prop wash.

Now he pushed the throttles forward and the engines responded beautifully. The runway markers seemed to be flying by faster and faster. He eased the control column back until the nose wheel came up, then he moved it back to neutral. She was getting light on the mains and ready to fly. He pulled back on the control column again and she began to lift off. Always there was that good feeling of breaking free, lifting, and climbing as the altimeter began to move.

"Pop the gears, Bill."

Soon he was making the first turn and entering the pattern for forming up into the boxes before heading out over the shore line of England and the Channel. It was beautiful down there with the patchwork designs of browns and greens of the Essex farmlands, but there was little time for looking. Too much work had to be done before the bomb run.

"Zola, what's she looking like back there?"

"Fine, Capt'n, everything's okay."

"We should be picking up our fighter escort soon. Nat, go back and clear the top turret guns. Pic, take care of the waist guns. We should be over the French coast in a few minutes. Hunsicker, get set to locate our IP."

Later in the flight black explosions of flak began to burst ahead. "There they are men, the coastal guns are zeroing in on our lead ships. Budge, we'll start evasive tactics until we hit the IP."

"Capt'n, it's really cloudy and tough to pick up the target."

"Hunsicker, if you can't pick it up then drop on Coiner's signal."

"Yes, Sir, it looks like 10/10 over target. We'll never get a good drop with this overcast. Bomb bays open, hold your heading Capt'n. I still can't sight the target ... There goes Coiner's drop ... Bombs away."

It all happened quickly and Quinn began to bank the ship in a sharp left turn, staying with the "A" box on its flight back to Rivenhall. The flak had been light and ineffective with no enemy aircraft spotted, but the cloud mass had been a factor in the ineffectiveness of the enemy anti-aircraft gunners, as well as the poor results of the bomb mission. Flying through overcast presented nearly as much danger as flying through a flak barrage since the chance of mid-air collisions was always a possibility destroying two or more aircraft in the disaster.

Upon landing and being debriefed the mission seemed to be a failure. Only eighteen planes had dropped their bombs. The other eighteen were flying in too thick cloud cover to make a drop at all. The photos over target were near useless so the mission would be classed as poor. The men were all disappointed, but it couldn't be helped. Their dejection could be felt in the mess hall that night. There was conversation but not much of the laughter and hilarity which usually was present.

Nat had discovered that their crew was on the loading list again for tomorrow and it looked like the loading list was identical to today's list. Maybe Coiner was giving them a message. Do it right or you'll have to repeat it or maybe he was just giving them a second chance. They really didn't mind since it was a chance to redeem the bad results of their first mission. "Capt'n, are we really scheduled for tomorrow's mission?"

"That's what I've heard, the loading list is identical to the one before."

"Is the target the same?"

"That, I don't know ... We'll have to wait until briefing tomorrow at 0700."

Sleeping was fitful that night for most of the men as they wondered about their assignment and had no answers. By morning chow there was the same inquisitive mood ... Where was the target? What was the cloud cover? But, most questions would be answered in a short while at the briefing.

The men filed into the briefing room and immediately looked at the map with the target area spotted by a red string which ran from Rivenhall to target. It was definitely in the Pas de Calais area as was yesterday's mission, but it looked like a different specific target.

Colonel Coiner had come to the platform with a determined look, "Men, we failed yesterday ... Our bombing results were poor. The weather didn't cooperate, but we must have good bombing results irregardless of bad weather conditions. You probably noticed our same loading list as yesterday ... It's no accident ... I want you men to show the Group we can turn bad luck into good with superior flying, navigating, and bombing skills. Our target is Bois De Coupelle, another V-1 flying bomb site. Write down these coordinates and emergency radio frequencies posted on the left. Our cloud cover is light with 2/10 cumulus, so targets should be easy to spot. Flak is expected to be light in this area. Fighter escort at 0820 ... Take off time is 0800 ... Let's get the job done this trip, men ... Load at 0745 ... Engine start up at 0750 and watch for the green flare."

The men filed out with a determination to hit their designated target. Every man was mentally going over his position's duties as they rode out to the hardstands. After take off, the navigators were carefully plotting their course to put the bombardiers in the best position for sighting the target after IP was reached.

The cloud cover was light as had been forecast and they encountered little flak except from the coastal guns as they crossed into France.

"Right 1 degree, Capt'n ... Hold on heading ... Steady now ... Bombs away." The ship felt like it was in an updraft as the load dropped and a lighter ship suddenly gained altitude. Quinn banked his Marauder into a tight turn and leveled off to join the others in return. Now Zola and Nat could see some of the bombs exploding down below.

"It looks like a good bomb drop ... All the ships are blanketing the same area, Captain," Zola called over the intercom.

The strike photos would tell the complete story. The men were confident at the debriefing sessions. Capt. McLeod was the S-2 officer asking all the questions.

"Lt. Daoust, did you spot a definite target?"

"Yes, Sir, it was identified correctly."

"Lt. Cook, how do you think the bombing classed?"

"It was good to excellent, Sir ... We saw a solid blanket on the whole target area."

"Lt. Budge, was there any enemy fighter activity?"

"No, Sir, not any that we could see from "A" box."

"How about flak?"

"Some, Sir, maybe medium to light."

"Okay, men, dismissed ... Good mission ... The chow hall is open if you guys are hungry."

Most of the men headed to the mess hall, now relieved that their mission had been successful. Their talk was centered on the bombing strike. "Did all of our planes drop on the target?"

"Yeah, there were only a few bombs brought back. I think one of the planes in "B" box had a malfunction on the bomb bays and couldn't drop."

"That Col. Coiner is a fighter isn't he?"

"Yeah, you have got to respect a group commander who leads his men into battle and takes the spot of greatest risk."

"I know I'm not jealous of his spot as lead aircraft ... That's the plane the German guns zero in on and if we get enemy fighter attacks, they head straight for the lead bomber."

"Yeah, those guys are smart ... They know how to disrupt a bombing run."

"Well, I'm not anxious to see any enemy aircraft, but if we do I hope our fighter cover will take care of them."

"Nat, what does the loading list look like for tomorrow?"

"The guys who flew today get a rest and the other half of the Group are flying tomorrow."

"Boy, I'll be glad to get a rest ... I'm going to sleep late tomorrow ... I don't care if I miss chow."

"Yeah, you and me too ... Those early wake up calls get tough after a while."

West's crew arose late the next day and had gathered at the Operations Building to get some information on that day's target. They had heard the planes taking off about 8 o'clock, as there was no way to sleep through the noise of 36 Marauders screaming for altitude. But, they wondered where the target was located. All the men on base were asking the same question. Nat had stopped to talk to some of his crew, "Lt. Budge, did they head for another V-1 site?"

"I think so ... The target was Vacqueriette, another Pas de Calais mission, but they said this place was tough and plenty of gun emplacements in the area."

"I hope those guys make out as well as we did yesterday."

"Yeah, but there is no telling ... What time will they get back?"

"Probably around 1130."

"Hey, it's nearly that time now ... Let's go down to the hardstands near the main runway and watch them come in."

"How many of our squadron went on this one?"

"About nine ships, I think, in our squadron and 36 total ... Barnett, Silverbach, North, Taylor, and Luther Williams."

"Capt'n, take a look past that cloud bank way out there. Is that some of our Marauders?"

"It looks like them."

Minutes passed as the ships got closer and the men were barely able to identify the squadron codes on them.

"Capt'n, look, some of them are breaking formation and shooting red flares and there are some with feathered props and engines smoking."

"Yeah, there must be some of them shot up pretty bad. Look at that first plane coming in! That poor guy has gotten nearly his whole tail blown off! It will be a miracle if he can land that thing. He's skidding down the runway and pulled off into the grass."

"It's a good thing 'cause here comes another one crippled worse than him. His right engine is on fire and he's coming in for a belly landing!... Here comes the fire trucks and the crash crews. This looks like a bad one. She's belly landing ... Oh my gosh, that thing is crashing on the side of the runway!" In just a few minutes it was over. The screeching, grinding, and tearing were over and the damaged plane sat burning on the runway. The crew were coming out of the waist hatch as crash trucks were helping them and shooting foam on the burning engine and cockpit area.

"I hope that ship don't explode ... Two more planes are landing and they may all go up in smoke if that baby blows."

"They're all shot up pretty terrible, Zola ... Those guys must have caught some real heavy flak."

"Yeah, here come some more on single engine and damaged in the tail section ... I don't see how those planes are flying."

"Look at that one with a section of the fuselage shot out. That's terrible ... How many have come in so far?"

"About ten or more, but I wonder now how many of our planes got shot down over France?"

"Well, we can walk down to the debriefing area and see if some of our pilots are there ... Maybe they can tell us what happened. Here come the rest of the Group. It looks like they are okay ... Maybe some shrapnel holes, but still flying."

"Hey, there's another red flare. The ship looks okay, but something must be wrong. Oh, Oh, here comes an ambulance ... There must be a guy wounded on that one. Why is it taking them so long? Maybe the hatch is jammed ... Yeah, there they go ... They are pulling him out of the waist hatch."

"Dog-gone, I really hate that ... These guys got shot up something awful. It looks like most of the planes made it in ... I've counted nearly thirty six."

"Let's go back to the debriefing area and find out if anyone in our squadron got hurt ... Maybe we can check with Lt. North or some of the other pilots and see what happened."

The debriefing area was in confusion as the men were trying to explain what happened on the mission to the S-2 Intelligence Officers. Quinn looked for a familiar face and finally spotted one of the pilots in his squadron.

"Hey, Bergman, what happened to you guys?"

"It was a tough one today, John ... Flak was everywhere, on the IP, on target, and after the bomb drop ... They were on us the whole time and they had our range perfect ... I'm surprised we didn't lose a few ships."

"Did they all get back?"

"Yes, as far as I know, all of us made it in. Our squadron didn't catch it as bad as some of the others. Some guy was wounded in one of the other squadrons, but I haven't found out who it was ... Boy, that flak was horrible ... Those German 88's can really work you over if they get your range and heading. It was so thick you couldn't keep the ship on course during the bomb run. I didn't have any idea it could be that bad."

"Come on, let's all go to the mess hall ... They have left it open waiting for you guys to come in."

"No, not yet .... I want to wait for my crew ... Some of them may be shook-up and want to talk about the mission."

"Yeah, I understand. I'll see you later."

Quinn walked to the mess hall with Budge and Zola. He was thinking that they might be a help to some of the others who flew today. They had been through a lot.

The afternoon brought some new concerns. The loading list would be posted soon and the men wondered who would be going tomorrow, or maybe the mission would be scrubbed if bad weather set in. There was a crowd standing in front to get a better look.

"Capt'n, we're on it and Col. Coiner is leading the first box again. That makes four straight missions for him ... The pilots in our squadron are Allen, Williams, North, Patterson, Quiggle, Silverbach, Ryherd, and Thompson ... Capt'n, there's some of these guys who flew today's mission .... Silverbach, North, Patterson, and Williams flew it and Bergman did too."

"One more mission for Coiner and he'll have the first air medal in our Group."

"Yeah, no one can say he's an armchair commander ... That guy is in the thick of battle everywhere we go."

The talk at chow that night was about the thirteen ships which had been damaged in the last mission. It was a high price to pay as the mission itself had been classed as poor results due to poor visibility. They would know more about it at the briefing the next morning. Some of the men hit the sack early as wake up call was set for 0600. Others stopped at the Aero Club to listen to the juke box or write letters home.

Wake up was called early and after morning chow the men made the short walk to the briefing room. Col. Coiner was going over some details of the last mission.

"A lot of you men were on yesterday's mission and saw the coastal guns take a heavy toll on our ships. Here's your chance to blow some of them away. We are going to hit the coastal defense guns at Benerville. All your coordinates are posted on the board. Take off at 0830 and let's put those German 88's out of commission."

The men were determined to make this strike a good one and it showed in every phase of the mission from take off to bomb drop. After the "Bombs Away" call from the bombardiers, the aircraft made their first turn and the men could spot the bomb pattern. It looked real good. They had laid a solid blanket of bombs down and they were satisfied. Back at the field, debriefing was short as Major McLeod was certain the mission classified as good. All bombs had been dropped on the target area. Very little flak had been encountered. No enemy aircraft had been sighted and none of their aircraft had been damaged. Col. Coiner would be happy.

Tomorrow would be their fifth mission and Col. Coiner was set to lead it with Major Dempster leading the second box. The loading list had included West's crew again, but late in the night the weather closed in with solid cloud cover and rain. By morning the men woke up to discover the mission had been scrubbed for the day.

There was nothing to do on those days but go to the Aero Club or get some extra sack time. Usually the men complained that they had rather be flying. Some just enjoyed the day off while others wanted to get their missions in as soon as possible and get back stateside. A rainy day was cheating them out of an extra mission that would help get them home sooner. Many times these days were used for training and classwork. Bombardiers needed practice with the Norden bombsight and pilots needed more training in ETO procedures. So it went with constant retraining and more class time especially as new men were constantly coming into the Group to fill empty places and build the units' strength to normal.

On scrubbed missions the loading list would carry over intact to the next mission, so the men would be on stand-by or alert until they had flown a mission. West's crew was still scheduled until the weather cleared and they flew their mission. The next morning was not much better with rain and 10/10 overcast, but the bombing had been scheduled anyway.

At the briefing Col. Coiner admitted that the weather was bad, but they had predicted a possible clearing over the target so the mission was started. There would be forty-one aircraft participating and formed up into two boxes. Coiner and Dempster would be leading. The target was another V-1 base at Bois Coquerel. They would carry eight 500 pound bombs each and safety precautions were stressed since most of their flying would be in completely overcast conditions.

After the briefing the men slowly filed out and began to throw their flight gear into the waiting trucks. It was socked-in and no good for flying. Even just taking off in this kind of weather was guess work. All were in trouble when they couldn't see the runway markers or the tree line at the end of the runway. The planes lined up and waited for the flare. After the signal they began to move out and cautiously, one by one, they headed out to the coast of France.

The overcast topped out at 6,000 feet so it wouldn't be too bad if they got a clearing over the target. It had cleared some by the time they reached the French coast and the German coastal guns were beginning to pound them with antiaircraft shells. It was a heavy barrage as the German gunners had picked up the range quickly. Flak bursts were everywhere as Zola called over the intercom, "Capt'n, we've taken some hits in the midsection."

"Okay, what's the condition back there ... Are any of the hydraulic lines cut?"

"No Sir, not yet. Not any of the crew are hit but that last close burst just blew some holes back near the left waist gun position."

The flak bursts continued to hit all around them as West called to Hunsicker on the intercom.

"How long until we are over the target, Hunsicker?"

"About 5 minutes, Sir."

Quinn turned to Budge and said, "Boy, this flak is terrible. It hasn't knocked any planes down yet but I know they must be getting hits because we are catching it so bad."

"Yeah, there's an engine smoking in one of the planes on our right."

"I hope he can manage to keep up on the single engine. She's smoking real bad now."

Hunsicker's voice came over the intercom, "Hold her on course, Capt'n, we're thirty seconds from target."

"Budge, we're in solid overcast again ... Nobody will hit the target unless we get out of this cloud cover."

"Hunsicker, what's it look like?"

"Too thick, Sir, they'll have to recall this one."

"Wait a minute ... Okay, Coiner has just recalled ... We're going to climb to 7,000 and try to break out of this overcast before making our turn back to base ... Give her a few minutes."

"All right, she's breaking out ... There are others making their turn ... We'll pick up our heading back and try to stay above the overcast to base."

"Captain, I hope we don't get a clearing below us when we cross the French coast again or those Jerrys will be shooting at us again."

"Yeah, me too ... Hold your breath ... We'll be there in a minute."

It didn't take quite that long as suddenly flak began bursting around their ships and they knew the coastal guns had gotten their range. As they moved out into a clearing from the overcast, the flak became more accurate and intense. Black puffs of flak explosions were everywhere as Quinn looked at the planes in the first box to see if any had been damaged. In an instant, his ship lurched sideways after a deafening noise.

Zola's voice came in loud, "We're hit Capt'n ... There's a big hole just forward of the waist guns, and some hydraulic lines cut. There's fluid all over the floor back here."

"Okay, check it out and see if she's damaged anywhere else."

The other ships were catching it also, as several had taken hits in their engines and were smoking badly. They had feathered their props and were trying to keep with the Group on single engine but it was no use, they were continually losing altitude. They began to straggle behind and West hoped they would make it to the field at Rivenhall. Some looked as if their engine nacelles were on fire as smoke billowed out of the engine compartments and all hoped their fires wouldn't spread to the wings or to the tanks and cause an explosion. They had extinguishers in the nacelles, but sometimes they wouldn't work and the crew was forced to keep on flying and watch the fire, praying that it would burn itself out and not reach any of the vital parts of the aircraft.

It was nearly an hour's flight time to Rivenhall so they nursed their damaged planes along and looked for the familiar coast of England and their flying fields in Essex. Minutes seemed to be hours until Budge and West saw friendly land on the horizon.

"There it is ... The coast of England ... Looks good doesn't it?"

"It sure does. Now all we have to worry about is whether we have enough hydraulic pressure to get our gears down."

"Zola, stand by to crank down the gears if we need it."

"Yes, Sir, I think they will make it down okay."

"All right ... Here goes." West was talking slowly, every step of the let down. "Slow turn into the pattern ... Fuel booster pumps on ... Mixture to auto rich ... Props to auto constant ... Push landing gear lever to down ... Slow her down ... There goes the horn ... Now the lock pins are in ... She's okay now ... Lower wing flaps ... Bring her in fast, slightly nose down, now flare out."

They heard the screeching of tires on concrete and they all breathed a sigh of relief. Then, Quinn taxied to the hardstand area and cut the engines as the crew began to pile out through the nose wheel well and walk to the waiting trucks.

Zola was talking, "Capt'n, I was holding my breath on those gears coming down."

"So was I ... The indicator light was on but it still was a good sound to hear those tires squeal on the runway ... What was the total damage?"

"Some pretty good size holes shot in the waist and I think those hydraulic lines that were cut go to the bomb bay doors, but it sure was a mess back there."

"Okay, the debriefing teams will want to know about it."

When they got there others were already in the process of filling out forms and answering questions. One of the S-2 officers saw Quinn's crew come in and sit down.

"Okay, West, how was it?"

"Well, we got some flak damage over the coast going in. The target was completely socked-in and we couldn't drop. No enemy fighters were sighted. And we caught some more flak damage going back over the French coast. That's about it, Captain. Did all of our aircraft make it back to Rivenhall?"

"Yes, as far as we know, there were sixteen ships damaged but none were lost."

"It was a tough mission, McLeod, to get shot at twice and still not be able to drop on the target."

"Yeah, I know ... That's the breaks. You guys are dismissed. There's hot chow in the mess hall ... Thanks Captain."

The men weren't too happy with the results, but they had flown enough missions to realize that some were winners and some were losers, and this was one of the lost ones.

That afternoon when the loading list was posted, Nat saw it first. He quickly headed back to West's quarters where Budge and West were talking.

"Captain, I can't believe it, but we're scheduled again for tomorrow."

"Yeah, maybe they're trying to get us home early. We'll have our fifty missions in no time if they keep this up," West laughed.

"I know, but I don't think they are doing us any favors by pushing us into those flak barrages every day without a let-up. I used to laugh at those guys who said they were 'flak-happy.' Now I know what they are talking about."

"Well, don't feel too bad about it. There could be some 48-hour passes around the corner to those crews having a string of missions to their credit."

"Capt'n, I believe you know something and are keeping it from us."

"No, but there may be a break coming ... Wait and see."

"Dog-gone, I would really like to see London."

"Yeah, we all would."

Another morning came with rain and near solid cloud cover. It was beginning to be expected. There were very few days considered to be good flying weather lately. It was no surprise to the men to find that the mission was still scheduled and a briefing had been set for 0710.

The target was a marshalling yard at St. Ghislain near Mons. Dempster and Captain Wood were leading. It was still raining and the mission was beginning to look like a carbon copy of the same weather they had yesterday. The planes took off and headed for the target only to find that there was 10/10 cumulus over the target area. Out of thirty-seven aircraft which flew the mission only five bombed the target, so it was classified as poor.

There was a bright spot in the day for Quinn's crew, however, because this was their fifth combat mission and they would be awarded the Airman's Medal. It would be weeks before they would receive the commendation, but at least they had already earned it. At chow that night the crew was happy about it.

"Hey, how about that Airman's Medal?"

"Yeah, at least we've got something to show for our efforts this past week of dodging flak."

Pic came into the mess hall looking for them.

"Well, you can chalk up one more 'cause we're scheduled to fly tomorrow."

"Oh no, what are they trying to do to us? See how long it will take to get us flak-happy?"

"Zola, you were flak-happy before we got here."

They all laughed. It was a way of letting off a little steam and Zola liked the good natured kidding and jokes.

The morning brought a change in the weather and at daylight they could see the clouds breaking with definite clearing and no rain. It was great. Now the crews could feel that they would have a good shot at hitting the target.

The briefing by Col. Coiner was even more optimistic. Col. Allen and McLeod were leading, and the meteorologist's report looked good. The target was a defended beach area near Ouistreham, all the way across the channel to the Normandy coast.

The men all loaded into the trucks with high hopes of getting a good strike at the enemy. Thirty-eight aircraft were going this time. The hardstand area seemed alive with activity as the planes' engines thundered to life and the aircraft began to taxi out to the main runway. The take off and form up went like clockwork. There had been so many missions lately which were uncertain because of the overcast. Now they could see their wingmen and it made a difference in their high spirits. This would be a longer mission than usual as most of their early missions had been close to the channel coast of France. Now they were flying northeast to the coast of Normandy.

The German high command had been expecting Allied landings in France for months now and preparations had been going on at a fast pace. The coastal defenses were enlarged with more gun emplacements, sea and land mines placed, and concrete and barbed wire obstacles strung along the beaches. The rail and road traffic had reached a feverish pace with trains and trucks hauling supplies to the defended areas as fast as possible. They knew the attack was coming but they didn't know where it would strike.

Normandy was not a militarily significant target to the American airmen at this time. It probably went unnoticed as another bombing of just another wide section of the coastal antiaircraft guns. But to the high echelons of the Allied forces this was all part of a strategy called "interdiction," which was fundamentally the attacks on bridges, roads, and defended areas to seal off a proposed combat landing area for Allied troops. This was done weeks and months ahead of the possible dates for the actual landings. In order to keep the German command unsuspecting of the probable landing site, the Allies bombed nearly twice as many targets in the Pas de Calais area as they did near the landing zone itself.

Later the world would know these landings as "D-Day" and this particular area where the bombing mission was heading would become the famous "SWORD" beach of that section of the Normandy landings.

The mission went well as thirty-seven aircraft hit the target area and 298 one thousand pound bombs rained down on the coastal target. All bombs were dropped ... None brought back to Base. The strike photos showed 97 percent of all bombs were in the strike zone. It was a tremendous achievement and a classification of very good from S-2 evaluations of the bombing effectiveness. The crews of the returning ships had seen the bombing blanket cover the area. They knew it had been a good mission. The best part of the mission was the light flak and no damaged aircraft to worry about on the way home. Quinn's crew were proud of the results but they also wondered if they would be scheduled for tomorrow's mission. Zola and Nat came to the flight deck with the same questions.

"Capt'n, do you think they will schedule us again for tomorrow?"

"Well, let me ask you a question ... Had you rather go to London for three days?"

"Man Alive, I can't believe it ... I thought you had something up your sleeve when you talked to me yesterday about our work load. I bet you knew it all the time, didn't you ... Tell the truth."

"No, I wasn't sure, but Coiner had talked about running some consecutive missions and then giving the crews a 48 hour pass ... But, we've got it if we want to take it ... Friday, Saturday, and Sunday in London. You'll have to pick up your passes in Operations when we get in."

"Missouri Mule II" and crew; Capt. George Parker, pilot; Lt. Moore, co-pilot; Lt. Cartmill, bombardier; and Sgts. Price, Brewer, Billings, Min, and Garvie.

Capt. Moses J. Gatewood and crew, 597th Squadron. Notice the crew wearing flak jackets and special helmets.

Capt. C. A. Thompson and crew, 599th Squadron. Crewmen are "Walking the Props" to clear oil from the lower cylinders.

Lt. Senart, pilot of "Bar-Fly" with his crew; Lt. Harlan, Sgts. Celeste, Bennedict, Nevin Price and Stiffler.

The "Susan J" with crew, pilot Capt. Smith, Lt. Blomberg, Lt. Stevens, Sgts. Curl, Windham, and their crew chief.

Major K. C. Dempster and a nine man lead crew; Lt. Sloan, co-pilot; Capt. Bero, Group navigator; Lt. Breen, bombardier, kneeling at right; Sgts. Wisotsky, Kuzma, and Powers.

"Innocence Abroad"; Capt. Buckler, Lt. Eggleston, Lt. Cotter, Lt. Cashin, Sgts. Zieliuski, Clark, Coxey, Lott and Berger. Last 3 are Group Photographers

The "Leapin Lena" crew; Sgt. Bunes, Lt. Norten, pilot Lt. Bergman, Sgts. Hansen, Neil McGinnis, Lt. Norton, Lt. Brooks, and Sgt. Silva.

James Wylie, a British war orphan, with a few of his "American fathers" who adopted him into the 598th Bomb Sqd.

CHAPTER EIGHT

THREE DAY PASS TO LONDON

The men were elated and before the plane had landed they were on the intercom to West and began asking questions, "Do we want a three day pass? I can't wait. When do we leave?"

"If you can hurry and throw some clothes in your B-4 bags, catch a GI truck to Chelmsford, and get the 4 o'clock train, we should be in London some time after 5:30 or 6:00 p.m."

"What about some of the other crews ... Will they be going?"

"Yes, I think Bronson's and Silverbach's crews are going, and maybe Barnett's crew also."

"Hey, that's great ... Donzello and Russell may go ... Also Sears, Gauker, and Kitrick. Man, we'll have a great time."

Quinn wheeled the Marauder into the hardstand area and smiled knowing he had a happy crew on board. They piled into the truck and headed for the debriefing, then made a hurried stop at Operations for their passes. With a quick packing of clothes, they were on their way to the main gate waiting for a truck to get them to Chelmsford's train station.

Pic yelled, "Here comes our ride."

"Hey, hold it for a minute ... Here come some more of the outfit. There's Donzello and Daoust and Barnett."

"Let's hold it a while before leaving ... There's some more coming. Yeah, it's Ray Snow, Roebuck, and Skarles ... Boy, we are going to have a party in London."

The ride to Chelmsford was alive with laughter and joking. The men were excited and happy to be going on a trip to London. But more than this, there was the great feeling of being away from the war for a few days. Zola and Pic were talking to Quinn.

"Capt'n, how many of the crews do you think were released from duty this weekend? It looks like we should have more going with us."

"Well, I think most of our squadron has been released for either 24 or 48 hour passes, but a lot of those guys probably want to stay on base and play baseball or football, get plenty of extra sack time, and get some good meals at the mess hall ... Maybe take in a dance at Braintree or Chelmsford ... Some of the men said they were going to South-end-on-sea ... They like it down there, but I wanted to see London. How about you guys?"

"Yeah, me too, Capt'n, it's a chance of a lifetime to see a city like that."

Quinn laughed, "We might not see all the city, but for the servicemen, Piccadilly Circus is where everything happens ... The Rainbow Club U.S.O. is there and most of the hotels and pubs that the military like."

"Where are we going to get rooms?"

"Well, you can actually get a bed at the U.S.O. ... Two blankets and a bunk for overnight, but most of the guys like the Imperial or Regent Palace or Royal Hotel ... Where we get off at Piccadilly Station will be close to everything."

It was only a short ride to Chelmsford and soon they were approaching the station.

"Hey, is that the train station?"

"Yep, that's it ... Nothing really big, but clean and functional."

It looked much like the typical small town rail stations of the 1930's back home.

"Okay, men, pile out and get your tickets."

Suddenly the shrillness of a train whistle broke into the hubbub of conversation, and the men went out to see the train come in. It was a slow process as the engine slowed and brakes screeched, stopping the cars at the station platform. The coaches all had compartments for six to be comfortably seated or possibly eight crowded in. There was a main isle and compartments on the other side of the coach.

The men came into the cars laughing and pushing for the seats closest to the windows. It wasn't long before the hissing of the steam signaled the brake's release ... Then with a few jerking motions the train began to pull out from the station. A whistle blast and they began rolling through the environs of Chelmsford, heading for the open countryside and the beautiful green meadows and brown plowed fields of Essexshire. They delighted in the scenery of cattle grazing on the hillsides and small farms bordered by the stone fences surrounding the farm houses and utility buildings. It would remind many of them of their rural upbringing and to all it was a pleasant change from the tents and quonset huts of army life.

The train was slowing at every crossing with a whistle blast to signal its approach. It seemed like a long time to the men before their train began to approach the outskirts of London, but soon after reaching the suburbs the train began to go slowly and several more station stops were made. Finally, they called Victoria Station and the men got out and made their way up the stairs to the street level.

It was not far to Piccadilly Circus where they would find hotels and entertainment centers. There were taxis at the station entrance, but usually the servicemen preferred to walk as everything was close.

"Capt'n, are you and Budge going with us tonight?"

"Sure, we'll meet you at the Rainbow Club tonight at eight and we can make some plans there for tomorrow. I'm going to try and get a room at the Regent Palace Hotel. Are most of you going to the Imperial?"

"We'll try there first but we'll meet you at eight o'clock tonight."

"Okay, see you guys then."

Quinn had packed more clothes than he had originally intended and his B-4 bag seemed over packed. He decided to get a taxi to the hotel. It was a little different from the taxis in the States, as the driver's compartment was open air behind a windscreen and the passengers seat was enclosed. It wasn't far from the hotel but he would enjoy the ride. He smiled as he thought of what some of the men said about London. They said that GI's start out eating steaks and riding taxis, then when their money gets low they start eating fish and chips and walking. That was pretty close to the truth.

The hotel was beautiful and well appointed especially when someone had just come from the barracks and military life. It wasn't long until eight o'clock so he hurried to get a change into uniform and make his way to the U.S.O.

He walked and found his way easily as he noticed a large group of servicemen going into the Club entrance. He looked for Zola first as he could usually see him in the center of a bunch talking with Nat, Pic, and the others. There they were ... Quinn could see a large group.

Zola shouted, "Capt'n, over here ... Come on over and sit down with us."

Quinn walked over to the group as Zola spoke.

"Capt'n, I want you to meet some great guys from the 322nd."

"All right, I'm glad to meet all of you ... You're at Andrews with General Anderson?"

"Yes, Sir, we're not far from your field at Rivenhall."

"I know, and I think we may have flown some missions together on combined operations."

They smiled, "Yes, Sir, they put so many planes up now it's hard to tell who you are flying with."

Zola spoke again, "Capt'n, this is Sergeant Houser Hall with the 451st Squadron."

"Hi, Houser, it's good to meet you."

"Same here, Captain. You've got a good bunch of men here."

"Yes, I think so ... What part of the States do you hail from, Houser?"

"Memphis, Tennessee, Capt'n."

"Dog-gone, I've got kin folks who live in Memphis, and my home is just fifty miles away in Sardis, Mississippi."

"Sure, I've been through Sardis. They're building a new dam down there."

"By golly, I can't believe we come from the same neck of the woods. There's an officer in our unit from Memphis, Captain Bronson, and we really have a good time talking about all the movie theaters and restaurants in town. The last time I was in Memphis my wife and I went to the Malco Theater and ate at Jim's Place. The next day we went to Overton Park and the Zoo."

"Captain, those places really bring back some memories. I'll be getting homesick if I think about it too much," Houser laughed loudly.

West asked, "Are all of you men in the flight echelons?"

"Yes, Sir, Houser here is engineer-gunner on the "Eleanor B" and Ray is a gunner on a ship they call "The Truman Committee.""

They all laughed. Truman was opposed to the B-26 and the men didn't like him because of it.

"Hey, you know we are going to have to name our aircraft one of these days," West replied.

"Well, Capt'n, we've all talked to Robbie about naming our ship, and since you don't use any swear words, we can't put anything bad on it. The strongest word we ever heard you say was 'By-Golly,' and Robbie said that would be a good choice for a name."

Zola yelled, "I'll vote for that. How about you guys?"

"Sounds okay to me," Nat chimed in.

"All right, Capt'n, how does it sound to you?"

"By golly, I never thought of naming a plane that, but it sounds good to me ... Okay, 'By-Golly' she is and when we get back to the field we can start dolling her up with a new name and paint on six bombs for the number of missions we've flown."

Zola stood up, "Capt'n, if you buy the cokes, I'll propose a toast to the best ship in the 397th. May she always come back from her missions."

"Yes," they all shouted their approval and laughed. They had just named their ship and it was a great occasion.

The talking and joking continued for a while until some of the men decided to go to the Pub at the Royal Hotel.

Zola said, "Capt'n, are you going to stick with us?"

"Well, it's getting pretty late for me ... I'll tell you what, if you guys want to, we'll get together for supper tomorrow. I know a good place for steaks ... It's the Grill Room at the Regent Palace."

"Sounds good to us. What time?"

"How about 8 o'clock?"

"Okay, we'll be there."

The main group headed to the Royal Pub, and Quinn walked to his hotel a few blocks away. It had seemed like a long day for him or maybe it was just the stress of the past week's missions finally catching up with him, but he welcomed the opportunity of going to bed a little earlier than usual. He left a wake-up call at the desk ... Then went upstairs and retired.

The next morning was overcast and cool. He had missed his wake-up call and had awakened at 10 o'clock. He dressed and had breakfast in the Grill Room, then made his way to Westminster Abbey. He was awed by the beauty of its structure and its seemingly perfect architecture. The houses of Parliament and Big Ben were landmarks of great beauty as was the Westminster Bridge over the Thames River. The Thames was a special attraction to him as he stayed for a long while on the Bridge and watched the boats come and go below him on the river. He began to think how much he wanted to have Ruby by his side now to see all these places of beauty and just to relax and enjoy each others company. Ruby always loved things of scenic beauty and he was sure she would delight in these scenes along the Thames. Maybe this was why he had been strangely drawn to this beautiful area because he was seeing it for Ruby as well as for himself. The Thames was a scene of perfect tranquillity with the occasional chiming of Big Ben. He walked through Hyde Park and enjoyed its beauty ... Then on to Buckingham Palace and the places of interest in the area. It took most of the day and before he realized it the day was far gone and time for his supper with the "By-Golly" crew. The name pleased him. How could they have chosen such a fitting name ... It just seemed to suit him perfectly. He thought about Zola's comment, "May she always come home." It was a comforting thought. Zola was a good man. He always could find the right things to say.

Quinn returned to his hotel and walked upstairs to get ready for the supper get together in the Grill Room. When he walked down to the lobby they were there waiting.

"Hey, you guys are a little early ... That's good. Come on, let's go eat. Are you all hungry?"

"Sure thing, just show us a menu."

The Grill Room was not crowded and the food and good conversation were just what they wanted. They talked about their sight-seeing trips and the different 598th squadron men they had run into last night.

"Capt'n, you won't believe it ... We ran into Budge and Daoust last night and so many others ... We wish you had been with us."

"Yeah, it sounds like you guys had a terrific time."

Nat asked Zola, "What's on the program for tonight?"

"Well, maybe a cinema or a look in at the Rainbow Club to see who's there tonight ... Hey, look at Captain West yawning ... I know he's going to hit the sack early," Zola laughed.

"No, not quite this soon ... But I would like to write a letter home while I'm here in London."

"All right, the meal was. terrific, Capt'n, what are you going to do tomorrow?"

"Well, there's still some more sight-seeing, but don't forget our train leaves Victoria Station at 4 PM tomorrow. Don't be late ... We need to get back to the field Sunday night. See you guys tomorrow."

"All right, Captain ... Thanks again for the swell meal."

"It was my pleasure, Nat. It was good to be with you all."

The men walked slowly through the lobby and Quinn went upstairs to his room. It had really been a good night, Quinn thought to himself. Those men were nearly like family to him ... Like brothers almost. They had been together since the early days at MacDill with more than a year of flying with them and now they were a team in combat. Men can't fly with a crew through flak and enemy fighter attacks without feeling a close bond with each other. They were bonded together by brotherly love and he knew they felt the same way about him.

It wasn't long before he had several letters written, but he was getting sleepy so he folded the letters together and went to bed.

The next day was again a little cool and cloudy, but not raining. It was Sunday morning and he wanted more than anything to hear a service at Westminster Abbey, so he made his way to the Westminster area. It was beautiful as he walked through the foyer and into the main Nave. Not a large crowd at all was there but it was a reverent one quietly listening to an organ prelude. The minister spoke and a choir sang very beautifully. It was a different service from his at home, but still it was wonderful to feel the presence of the Spirit in this beautiful house of God ... How lofty were its ceilings ... How exquisite were its wood and marble carvings. It was good to be in God's house wherever it might be. He left refreshed and uplifted as he heard the bells of several churches peeling their deep tones of music inviting all to come and worship.

The day had brightened somewhat as he walked to several tourist attractions. He wanted to walk again along the banks of the Thames, and then go early to Victoria Station for the train to Chelmsford. He wondered about the Group and where they had been bombing while he was away. It would be good to get back. He had been refreshed by these few days off ... Now he was ready to get back to the job of helping to win a war.

The station area was crowded mostly with servicemen returning to their units. He looked for the departures and their track numbers on the large boards at the station. There it was, Chelmsford ... 4 PM departure ... Track 9. He still had time before boarding so he sat on a bench near the track area and looked at a London Daily News. The war effort was going well ... Patton was rolling through Italy. Montgomery was making great gains, and the bombing effort was closing railroads and bridges everywhere in France.

It was May of '44 and Quinn could tell that something big was brewing. There was talk about Allied landings in France. It had to come sooner or later. The only question was when. The bombing missions looked like they should be pointing to landings somewhere, but where? He was deep in thought when he heard a familiar voice.

"Hi, Captain ... We thought we might find you today around Westminster, but we must have missed you."

"I was there ... We just didn't make contact. Did you guys get to see all the sights?"

"No, but we tried ... There just wasn't enough time."

"I know, that was my problem. I kept running out of time ... By the way, our train leaves in twenty minutes ... Maybe we had better go now. It will take a while to walk down to the tracks."

"Okay, we're ready to go."

The train ride was uneventful and they dozed in the coaches as time slipped by. Then suddenly the Train Master was calling the stop at Chelmsford before they realized it. Some of their buddies had knocked on the compartment window and woke them. The Army Air Force trucks were waiting ... It was a regular run for them to pick up the airmen on nights they were returning to base.

The trucks rolled into the headquarters area and unloaded.

"Pic, do you think the mess hall will still be open?"

"I don't know, but Sarge will get us a hot cup of coffee and maybe a few do-nuts."

"You know what I want to do more than anything else is to see some of the guys and find out how the bombing came out while we were gone."

"Yeah, me too ... Let's go by the Aero Club after we sign in and maybe we can find someone who knows."

They went to the Club after stowing their bags in the barracks. The crowd was there and it wasn't long before their conversations turned to the missions flown on the previous days.

CHAPTER NINE

COMBAT MISSIONS MANTES GASSICOURT TO ROUEN

The Aero Club was always a place for lively conversations about past missions, and Sergeant Bill Henry was giving them all the lowdown.

"Man, we had a real bad one today. We went to a V-1 bomb site and the flak was horrible. Eighteen planes were damaged and one guy was wounded. It was at Lottinghem. The two days before we were supposed to bomb a marshalling yard at Mantes Gassicourt and the weather was lousy. Both missions were recalled over the target and everyone brought their bomb loads back ... Hey, by the way, you guys are scheduled for tomorrow, probably for the same target. I hope you have a good run so we can finish that place. Looks like the weather is clearing so you should have a good mission."

"Well," Pic exclaimed, "we're back in harness again and ready to work."

The predictions were right ... They were going to the same place again, the marshalling yards at Mantes Gassicourt, but good weather allowed them to return with better results than before. The bombing was classed as good with 148 one thousand pounders hitting the target ... No damage to their aircraft and no enemy fighters sighted.

On May 2, 3, and 4, West's crew were not scheduled to fly and it was a time for baseball, trips to Braintree and Chelmsford, and a special time for painting the name "By-Golly" on their aircraft. One of the ground crew in the squadron did most of the art work for the men, and he had agreed to paint the name on their ship. Quinn and some of the crew were watching him work.

"Captain, how do you want the letters, all in block or in script?"

"Put it in script in big letters."

"Okay, anything else like a cartoon picture?"

"No, but I would like to have the names of my wife and son next to the left window of the cockpit area by my name."

"Sure thing, Captain, I'll put their names in block."

"Okay, print them, 'RUBE' and 'JOHNNIE'."

Quinn watched him as he worked, amazed at the man's dexterity. In a few minutes it was lined out and he had drawn the letters. All that was left was the actual painting.

West was pleased, "You really do good work. Did you have some art courses in school?"

"No, Sir, I just picked it up on my own."

"Then it's a natural gift ... Keep working and developing it."

"I plan to, Sir."

"Well, we really appreciate you doing this for our crew."

"You're welcome ... You probably know the bomb silhouettes are stenciled on, so every time you get a few more missions, let me know and I'll paint them on for you."

Quinn stayed for a few minutes after the painter had gone. It was just what he wanted ... Ruby and Johnny would be proud of their names on the aircraft. Not many would understand that "RUBE" was his pet name for his wife, but that suited him fine. Now Ruby and Johnny could fly beside him on his missions. It would give him a feeling of their nearness while he was flying.

The days were racing by and on May 8 they were scheduled again. Quinn stood at the bulletin board at Operations checking all the pilots names in his squadron who were flying the next day when Nat came by with several buddies.

"Capt'n, are we scheduled for tomorrow?"

"Yes, we're finally on the board ... I thought they had forgotten about us."

"Sir, I want you to meet some of our new men, Sgt. Mitchler and Sgt. Neil McGinnis."

"Well, I'm glad to meet you men. I've seen your names on the squadron roster and met the pilots on your crew, Steere and Lipscomb, they're good men."

"Yes, Sir, we think they're the best. We're glad to be in the 598th Squadron ... It looks like a good outfit."

"We're glad to have you, Neil ... Let us know if there is anything we can help your crew with."

"Yes, Sir."

West walked back to his quarters to finish some reports he had been working on and then he planned to hit the sack early. Tomorrow could be a rough day.

The briefing began at 0730 the next morning and Col. Coiner was going over the mission's details.

"Our target is a railroad bridge over the Seine River at Oissel. Berkenkamp and McLeod will be leading. Weather over the target area looks good. Flak is expected to be heavy. You'll be carrying thousand pound GP's fused for good penetration. Men, this is an important rail artery across the Seine. We've got to knock it out of operation. Start engines at 0800 ... Take off at 0810. Good luck."

The trucks were waiting to take the men to their aircraft. It would be only a few minutes until take off as the men jumped out of the trucks and went to their aircraft. West and the other officers had gotten there earlier and were talking. West spoke, "Daoust, we're glad to have you and Brooks on the run today. Barnett says you're the best bombardier in the Group."

"Thanks, Captain."

Fred Daoust had been with the original Group at MacDill and had flown with Barnett's crew during the early missions. West knew he was a good man; accurate, well trained, and quick to pick up a target in the Norden bombsight. He was an asset to the crew, a man of few words, but an anchor of steadfastness when the going got rough.

The men loaded in through the front nose wheel well. There were eight men in the plane today, a larger than usual number, but West was flying deputy lead and backing up McLeod. If anything happened to McLeod on the mission, West would take over as box leader, a great responsibility since the performance of the entire box of eighteen aircraft would depend upon him and his crew. Lt. Daoust's accuracy as the bombardier would be essential to the success of the mission.

Engines began to fire, starting with a cloud of exhaust, and the noise was deafening, but it was a good sound to the crew chiefs and the ground crew. They stood by to watch the action, proud of their individual ships. The seeming confusion of the eighteen aircraft thundering to life was like a tonic to them. They watched as the aircraft taxied out in line and one by one roar over the end of the runway. It was an exhilarating feeling.

Robbie was talking to some of the other ground crewman.

"Let's walk over to the runway edge and watch them take off."

Some of the planes had already roared by as a crewman asked Robbie, "Has West gone yet?"

"No, not yet, I'd know that plane nearly with my eyes closed. West is a fine pilot, and he's a real pro at getting the most out of his ship. Some guys might call it dangerous flying, but it's really cold calculation and knowing his Marauder. Watch him pop those gears early so he can squeeze every foot of altitude he can out of that overloaded Marauder."

"Hey, Robbie, here comes one giving it all she's got. Boy, listen to those engines."

"That's him ... Watch those gears ... Look at 'em ... Hot dog, that's West all right ... Look for the squadron code U2C ... That's him for sure. I told you I could tell when that plane flew over. When you listen to the engines about midfield they will be in 'sync' for a long stretch, then they'll break into the damdest thunder you've ever heard when the plane crosses the runway's end. It sounds like the 'Jugs' are going to burst, then that baby levels out and starts a slow, climbing turn. It's beautiful." The men watched the rest of the aircraft roaring across the runway, what a sight to behold, it made a man proud of everything America stood for to see that 'Might' being launched against the enemy.

The planes gained altitude quickly and began to circle and form the box patterns they would use to bomb the target. They were to meet the fighter escort at St. Catherine's Point at 0820. West saw them first and said, "There they are, off to the right at 2 o'clock high." They were a group of Spitfires today and they looked good with the English rondels on wings and fuselage. The flights droned on until they reached the French coast and began a pattern of 90 degree turns left then right, then left again in order to keep the German antiaircraft gunners from getting accurate range and heading on them. The flak was not very heavy today and the zigzag course was helping to keep them away from some of the worst concentrations of flak.

"Brooks, give us a position reading." West commanded.

"We're on course, ten minutes from IP, Sir."

"Daoust, are you pulling in some landmarks leading to the target?"

"Yes, Sir, we're on course."

There was increased tenseness as they drew closer to the target. This was the focal point of the whole mission. If things went wrong here it could turn into a missed target and a useless run.

"Captain, we're over the IP, hold her steady ... Bomb bay doors open ... You're on P.D.I."

The tenseness was building by the seconds as Budge grimaced and said, "Here comes the flak and it's heavy."

The flak was one of the heaviest barrages they had experienced. Black puffs of exploding shrapnel were everywhere. The worst part was the Germans evidently had their range and heading as the bursts were right in the area of both boxes and several planes were smoking already.

"Budge, look to the right, there goes one of our ships, and she's circling left with both engines smoking."

"Zola, watch that plane going down and count the chutes if you can."

"Captain, we're nearly over target. Hold her steady now. I've got the target in my sights. Hold her, BOMBS AWAY."

"Did anyone see chutes come out of that downed aircraft?"

"Yes, Sir, there were six ... Pic counted them. We couldn't identify the ship when she started down. I think it was in the other box."

Quinn was making his turn back to base when he felt it. A close flak burst and the plane seemed to swerve to the left.

"What happened back there, Zola?"

"I think we got a hit in the rudder."

"Okay, it sounded close ... The controls seem to be all right, so it must not have cut any control cables ... Hey, Zola, I feel some looseness in the rudder response ... Climb in the top turret and take a look when I move the rudder."

"Capt'n, there's a big piece knocked out of the lower part."

"All right, we better go easy on the rudder control until we get back to Station. Budge, there's some more planes smoking over on the left. I know there must be plenty of damage to our flight."

In a few more seconds they had flown out of range of the German guns. But soon they would be faced with the flak from the coastal batteries, and any other German antiaircraft emplacements along their route which got their range and began shooting at them.

Flak was really something fierce to contend with. No wonder men got "flak-happy" as the guys called it. They began to fear it and it became a psychological fear, a neurosis that made them afraid to fly. It was a nervous condition that required psychiatric treatment. Flak was an insidious thing that no one could do anything about, except to keep flying through it and hope none of it hit. It was something they were powerless against, just to sit there and wonder if the next shell would have the plane's number on it and they would become another number on the casualty lists.

The flak was light and ineffective over the coast. It was a good thing because the Marauders were damaged badly as it was. To have to take another flak barrage would have been the end of some of the stragglers with damage and wounded aboard. There had already been one man wounded in one of the ships. Also there was a badly crippled Marauder trying to keep with the Group, but continually losing altitude and distance. These damaged aircraft would shoot red distress flares from their planes and take precedence in the landing pattern, making sure that crash crews were there on the runways to help with emergencies. Other damaged aircraft followed in the pattern and the men in the control tower could assess damage to wings, fuselage, and engines of the landing aircraft.

Finally, the biggest part of the planes had landed and had taxied to their hardstands. Their waiting ground crews were inspecting damage and asking the air crews how it was during the mission.

"How was it, Capt'n?" Robbie asked.

"Bad, Robbie, the flak got us several times."

"Yes, Sir, I see the rudder ... How about the engines?"

"They came out okay ... Do you know who it was that crash landed out there a few minutes ago?"

"No, Sir, but some of the early crews said Lt. Freeman got shot down over the target."

"Yeah, we counted six chutes."

"They probably all made it out okay."

"I know, but I worry about what those guys will face in German prison camps."

"Yes, Sir."

"Check her over good, Robbie ... She may have to go to the maintenance shops with that rudder."

"Captain, I'm glad you and the crew got back okay ... We're always pulling for you when you're on a mission."

"Thanks, Robbie, we appreciate it."

The trucks had come to take them to the debriefing. It would be a long one because there was plenty to talk about. With one ship crashing in enemy territory, another crashing at the field, and twenty-six aircraft damaged, it had been a bad day for the Group.

West walked to the mess hall and quietly ate. He decided to write a few letters and then possibly turn in to bed a little early. They were not on the loading list for tomorrow so he could sleep a little later than usual.

The following day dawned fair and the mission was off to another V-1 bomb site in France near Le Grismont, but Quinn's crew had not been scheduled. So they spent most of the day playing baseball and relaxing at the Aero Club. When the men came in from flying the day's mission, everyone was interested in hearing what had taken place.

"Hey, Neil, come over and tell us what happened today."

"Well, we didn't catch any flak, but some of the other crews did ... Seven planes were damaged and a guy was wounded ... There was a bombardier killed on the bomb run ... I think it was Lt. Evanick ... He was flying with Lt. Thompson."

"Man, that's terrible ... I sure hate to hear that."

"You guys are on the schedule for tomorrow and so is our ship, so we'll be flying together."

"I wonder where we'll be going?"

"Don't tell anyone I told you but the rumor says a marshalling yard near

Creil is the target."

"McGinnis, how do you know all this stuff? They won't tell us anything."

"You've got to have pull, son, and it don't hurt to know a few people in Operations.

"The rumors had proved to be correct as Col. Coiner told them where the mission would be during the next day's briefing session. It was supposed to be an easy mission with little flak expected, but sometimes these easy runs turned out to be the most difficult.

This one proved to be just that with heavy flak over the target and nine planes returning with damage. Quinn's plane didn't get any damage on this mission and the crew was thankful for that.

The next day they were scheduled for a mission to bomb an enemy airfield at Beaumont Le Roger on the 11th of May. This was fortunately a successful mission with little flak and no aircraft damaged. It was classed as good with 406 bombs dropped on target.

On the 12th they were scheduled again to bomb coastal defenses at Etaples near Boulogne, France. Again it was another mission of little flak and no damage to their aircraft. But their luck was due to change for on the 13th of May, they were scheduled for their fourth straight day to bomb coastal defenses at Gravelines near Dunkirk.

A definite change in the type of targets chosen for the missions was noticed. At first they had concentrated on the V-1 flying bomb sites and marshalling yards, now they were centering on coastal defenses and railroad bridges. Historians now know that the plan was the "Interdiction" phase of weakening of German defenses prior to the Allied landings on "D-Day," June 6. However, the men flew the missions and could only guess at the possible overall plan of why these targets were of such great importance.

Col. Coiner was emphasizing this importance one morning, "Men, this is one of our most important missions to date. We are attacking a coastal defense line near Dunkirk. Each aircraft will carry two two-thousand pound bombs ... That's enough to tear some big holes in their defense lines if we place them well. Col. Allen, Wood, and Rhodes will lead the three boxes. Weather over target is clear ... Mark down your coordinates and times over IP and target. Fighter escort at 0820. Take off at 0800. Good luck, men."

The crews walked out to the trucks with a confidence born of believing in a strong man like Coiner. He was never afraid to lead them into battle and they knew it. Now, administrative duties had kept him on the base much of the time, but his thoughts were of the crews in the airwar. The men admired him for this and they would always do their best to bring honor to their unit and their country.

The trucks moved down the perimeter bordering the runways to the area where the squadron's Marauders were parked, waiting for their crews. The men unloaded and swung into their positions in the planes with pilot and copilot forward on the flight deck, then navigator and radio operator in the compartment behind them. The bombardier was situated in the nose forward of the flight deck and gunners at two waist guns or top turret and tail gun positions. Sometimes there would be a six man crew and other times a nine man crew depending on the situation. The lead crews were larger and carried more equipment.

The noise of engines began to add to the apparent confusion during take off time, with the supply and gas trucks moving from place to place and shouted instructions for last minute preparations. It was a hodgepodge of heightened activity, as the first Marauders began to pull away from the hardstands and taxi down the perimeter. In a few minutes they would be poised to push their throttles wide open and grab for altitude. Later the formation would set its course for the target and men would wait and watch for flak or enemy fighter formations.

The flight was relatively quiet until just before they reached the target, then suddenly flak began to explode everywhere.

"Daoust, this flak is getting worse ... How long to target?"

"Twenty seconds ... Hold her steady ... Here comes the target ... A few more seconds. Here she goes ... BOMBS AWAY!"

West watched the others in his group and swung left with them as they pulled away from the target."

Daoust, you must have put them on target. There's a line of explosions that looks good. Hey, Budge, look at that next box. Someone has gotten hit bad. It looks like Col. Wood's plane ... He is losing altitude and fighting to control that thing ... Watch for chutes and count if they start leaving the ship."

"It looks like he is making it okay, but losing more altitude. I hope he makes it to the field."

The flights droned on until they saw the English coast. They had lost sight of Col. Wood's plane and wondered what had happened to the crew. The field at Rivenhall slowly came into view and it always was a beautiful sight to know that they were home and had finished another mission. There were some damaged aircraft that had dropped out of formation and landed first. Then the others circled and landed one by one until all were down.

Quinn turned off the perimeter and braked at the squadron area. Quickly he cut his engines and started the process of his crew's emergence from the aircraft. He would be the last one out, but he was anxious for news about the crippled aircraft and wanted to know if it had been Col. Wood's plane.

"Robbie, has all of the 597th Squadron landed?"

"I don't know, Captain, I didn't count this time."

"Okay, here comes the jeep ... I'll ride to the debriefing and some of the men there will know."

The debriefing room seemed alive with the hubbub of voices as intelligence personnel were talking, asking questions, and the officers and the crews were giving their accounts of the happenings of the mission. Quinn spotted McLeod.

"Captain, let me ask you something."

"Okay, West, what is it?"

"We saw what looked to be Col. Wood's aircraft hit over the target and losing altitude ... Then we didn't see his aircraft landing. Has anyone reported seeing his plane?"

"We just got a radio report from West Malling in Kent. Wood crash landed there. Three of his crewmen were wounded and Lt. Evans was killed."

"Oh, my gosh, that's terrible ... I knew that ship was hurt bad when she first got hit."

The statistics didn't look good for that mission ... Three men wounded, one man killed, twelve planes damaged and one aircraft crashed in England. The bombing results were good, but it could never replace the loss of good men.

The "By-Golly" crew were given some days of rest and they took advantage of them by playing baseball and lounging at the Aero Club. Some nights they were released from duty and they went to some of the close towns for U.S.O. dances or to some of the favorite pubs for entertainment. At Braintree there was a favorite pub called the "White Hart." The airmen from several fields in the area would gather there to talk and find out what the other outfits were doing. There were servicemen from Andrews Air Base and Earles Colne, as well as Rivenhall and others.

While the crew were resting, the 397th was still busy bombing coastal defenses at Denain/Prouvy, Etaples, and St. Marie Au Bosc. Then on the 20th of May, West's crew were scheduled for the second mission that day at Varengeville Sur Mer near Dieppe. It was another coastal defense target area. The flak was light and all planes returned without damage a near perfect mission.

Quinn's crew flew a mission on May 22 returning to the coastal defenses near St. Marie Au Bosc. This was nearly a carbon copy of their previous mission with results good, no flak, and no damage. The men of the 397th Bomb Group had flown their last six missions without any damaged aircraft, but their good fortune was changing as their next mission scheduled for May 24 was to the widely protected harbor at Dieppe, it was later described as one of the hottest places they had ever flown over.

The briefing session was set for 1720 and as the men entered the room their eyes went immediately to the large map and the red string leading to the target ... It was Dieppe. One of the best fortified harbors in France. The men had heard stories of missions the other groups had flown to Dieppe ... It was a tough place ... Few missions ever went there without being shot to pieces by the many antiaircraft gun emplacements in the harbor area.

Colonel Coiner was explaining the mission slowly and with a seemingly deeper seriousness than he was usually accustomed.

"Men, this is one of our more important missions. The harbor at Dieppe receives thousands of tons of shipping daily to supply the German war machine in France. Our job is a big one ... To stop this flow of war material. The flak is expected to be heavy ... Use all your defensive tactics to avoid it until you make the bomb run. Fighter escort at 1820 at St. Catherine's Point. Weather over the target is clear to 2/10ths cumulus. Berkenkamp, Taylor, and McLeod are leading. Take off time at 1800."

The men filed out to the truck for a ride to the hardstands. West was talking to Daoust, "What do you think about this late evening mission?"

"It may fool the German gunners, but it may make it harder to pick out the target on the bomb run."

"Yeah, that's what I was thinking too ... Cloud cover could make it worse ... We'll just have to wait and see."

The ground crews at the hardstands were working until the last minute to get the aircraft ready ... Robbie was directing the work crews. "Hey, Smitty, get that gas truck to top these tanks off."

West asked, "Robbie, is she ready to go?"

"Just about, Sir, they gave us a hurry up job on this mission."

"Okay, men, let's load in ... Come on, Nat, Pic, Zola ... Daoust, you go on in and Budge and I will follow you."

The engines sounded good as they taxied out to the runway. Quinn had a good feeling about this mission. He knew it wasn't good to let feelings have a part in any mission, but it couldn't be helped ... Sometimes he felt good about a mission and other times there seemed to be numerous doubts about the outcome.

All went well until one minute to target, then it felt as if every German gun in Dieppe was zeroing in on the three boxes. Flak was everywhere.

Planes were getting hit, then going on single engine and struggling to keep their place in the box formation. West could see several hit badly desperately struggling to right their aircraft and continue on the bomb run. He heard Daoust call, "BOMBS AWAY," and he wheeled the plane hard left and tried to escape the pounding flak, but it was useless. It would be minutes before they would get out of range and clear the Dieppe area. These were the times that seconds felt like hours, but finally they were clear as West glanced at the rest of the Group's aircraft. A few planes were straggling with smoking engines, trying to make it back to England. Others were damaged by flak, but were managing to keep in formation. It would be a long trip back.

After the landing at Rivenhall, the men were anxious to hear of what had happened to the damaged ships. It wasn't long after the debriefing until news would get out from the intelligence sections.

Lt. Thompson's aircraft had been hit in the main fuel tank and right tire ... It took some expert airmanship to bring it in.

Lt. Gross had gotten heavy flak damage to his plane on the bomb run. His copilot, Lt. Chadbourne, was seriously injured and they had staggered back to England, crash landing at Newchurch. There were nine planes in all, damaged by flak, but West's plane had received no damage on this mission.

The men were elated as Zola and Nat left the plane slapping each other on the back.

Zola shouted, "Fifteen missions ... Sweat 'em, Nat, sweat 'em."

"Yeah, we've got to paint some more bombs on the plane."

"Capt'n, when do we get our ten days flak leave?"

"Maybe after about twenty missions if the Group is not on a big push to bomb extra targets."

"Okay, I'm ready to go right now ... How does London sound?"

"Well, it's your choice, but I'm thinking about Scotland. They say it is beautiful and nearly untouched by the war."

"All right, Capt'n. It's Scotland, if we can sweat five more missions."

The next day was a rest day for them, but the other planes in the Group had bombed a bridge at Liege, Belgium, about nine miles from the German border. There were twelve planes damaged. Col. Coiner and Col. Allen had lead the mission.

On May 26, West's crew were scheduled to bomb a Luftwaffe airfield at Chartres, France. The bombing results were poor due to weather conditions, but they had returned with four damaged planes in the Group.

They were scheduled the next day for a mission to bomb a railroad bridge at Lemanoir, and would have a new navigator on board, Lt. Douglas Cramer. For the last several missions, Lt. Daoust had been acting navigator-bombardier, and they would all be glad to have an extra man to take care of the heavy work load. Before the mission, Quinn had a few minutes to talk to the new navigator.

"Lt. Cramer, we're glad to have you with us."

"Thanks, Sir."

"You were with Captain Steere's crew, weren't you?"

"Yes, Sir, we've flown together since we joined the Group."

"Steere is a good man ... Come on up and meet some of the other crew members."

The mission turned out to be a good one with excellent bombing results and only five damaged aircraft. West was pleased with the bombing pattern and he felt that Cramer's navigation and Daoust's good bombing had made it all possible. Upon arriving back at the field, Quinn turned the plane into the hardstand area and the men began the unloading process. When West got out of the plane he noticed Robbie had a dog beside him. It was an English Spaniel marked black and white. It immediately reminded Quinn of a dog that he had in Sardis as a pet years ago.

"Hey, Robbie, where did the dog come from?" "'I think he followed Cramer over here. He's fine looking, isn't he?"

"He sure is. Come here, boy, come on." Quinn knelt down to pet him. "That's a good boy, hey, you like to be petted, don't you? Wait a minute now, don't lick my face. You're something big boy. You're okay."

"Captain, I believe that dog likes you."

"Well, you know, he reminds me of a dog I had a long time ago and I sure miss being around dogs."

"It looks like you've got a friend now ... That dog hasn't stopped wagging his tail ever since you petted him."

"Did you feed him?"

"Yes, Sir, one of the boys got some scraps from the mess hall."

"Good, keep him around here if you can. He'll probably sleep around the armorer's tents."

"He don't look like he wants to leave here."

"Robbie, I'll see you tomorrow if we are scheduled."

"All right, Sir."

Quinn got in the jeep and drove to the Operations area. They were scheduled again. This was three straight since they had their last day off. It seemed strange how the missions came in bunches, then they would be released for a few days, then back to the job of flying missions again. "Oh, well," he thought to himself, "that was life." He drove to the officer's quarters and decided to write a few letters and hit the sack early.

Briefing the next day was at 0800. The target was a railroad bridge at Lieges. It looked like they would hit every bridge on the Seine before it would slow down. Coiner had said nearly the same thing a few days ago. When he told them the overall plan was to neutralize the traffic on the Seine and the bridges above it.

The mission was a successful one with the bombing pattern classed as excellent. Seven planes had been damaged by flak, but there had been no casualties. Quinn pulled his plane into the squadron hardstands and the crew unloaded to find the black and white dog wagging his tail excitedly as he watched the men come out of the plane. They all stopped a minute to pet him, and as Quinn got to the group he said, "Looks like we've gotten a crew mascot."

The dog barked and Quinn reached down to pet him and 'roughhouse' with him a while."Have you guys named him yet?" Quinn asked.

"Some of the ground crew call him "Jiggs" ... They say when he sees our plane coming in he wags his tail so hard it looks like he's dancing a jig."

"All right, Jiggs, you've just gotten your new name. How is that, boy?" Quinn was playing rough with him, "How do you like that name fellow?"

The men were laughing with Quinn and Jiggs was loving the attention he was getting from all the crew. They had found a mascot and he would help take some of the tension away from their flying duties when they could reach down and pet a friendly dog.

That afternoon the men gathered at the Operations' bulletin board and something they had been waiting for was posted on it.

"Hey, Nat, come here and take a look at this."

Natanek read it aloud, "By order of Headquarters Ninth Air Force, General Order No. 141, 18 May 1944, the following personnel were awarded the Air Medal for meritorious achievement while participating in aerial flight in the ETO, each having participated in the required number of operational sorties against the enemy. Seventy five officers and enlisted men listed below."

"How about that, Buddy, we got our Air Medal. Did you see the bad news underneath it? We're on the loading list again. This is the fourth straight. If we keep this pace we'll have three oak leaf clusters to go on those Air Medals."

"Yeah, let's go to the Aero Club and see who's over there. I think there is a good movie tonight. Most of them are so old we've seen them a hundred times."

The next morning they were all gathered in the briefing room and Captain Steere's crew was with them. Zola and Nat were talking to

McGinnis and Schubin.

"Neil, I didn't think you guys were flying today."

"Yeah, they got us. Did you hear about the evening mission yesterday? Twenty-one planes got damaged by flak. It was the Maisons Laffitte railroad bridge in Paris. That must be the toughest place in the world for flak."

"You can say that again. Look at the red target string. We're going right back to Paris today."

Col. Coiner walked into the room and the men stood at attention.

"At ease, men. We've bombed nearly every railroad bridge there is across the Seine from the coast to Paris, and we have had some tough ones. Today does not look much different. It is the Conflans railroad bridge in Paris. Flak is expected to be heavy. I don't have to tell you how important it is to put all your bombs on target. There are French civilians and centuries old architectural structures that we don't want to get the reputation for destroying. Take down all your coordinates. Take off time is 0800."

The men were talking about flying over Paris, but they realized the flak was going to be rough.

Neil was talking as they walked out, "Zola, I've always wanted to go to Paris but this is the wrong way, son, we're going to have some German gunners for a welcoming committee."

"Yeah, I know."

The dire predictions proved to be correct for there were seven planes damaged by extremely heavy flak and sadly one of the Group photographers was killed by shrapnel from a near flak burst. They had placed their bombs well as the results were excellent with all bombs falling into the target area.

On May 30 they were scheduled again to bomb a highway bridge in the Meulan area close to Paris and along the Seine River. It was another tough mission for heavy flak concentrations as ten of the planes returned with damage. The bombing results were excellent. They were beginning to be a seasoned Group with a record of good bombing results even though flying through the thickest of flak.

The next day the "By-Golly" crew were scheduled to fly their 6th straight mission in as many days, and needless to say, their spirits were a little low. Nat and Zola were talking before the briefing.

"Boy, this is about to wear me out. When are they going to give us a breather?"

"I don't know, but I'm sure ready for one. Neil, your crew has been flying with us nearly every mission. This is awful."

"Yeah, I know it. What causes the problems is those days we fly two missions. That takes away our days off."

Col. Coiner came in for the briefing as the men stood at attention.

"At ease ... We have had a tough time with double missions these last few days, but I can promise you they are needed. There may be more before it gets any better, but I know I can count on you men to keep going ... You may be glad to know our title is now official. We are now known as the 'Bridge Busters.' I had it put in the records at Wing Headquarters." The men applauded and the briefing continued. "The target is a highway bridge along the Seine at Rouen. Weather over the target is poor with 6/10th cumulus. McLeod and Bronson are leading ... Write down all your radio frequencies and coordinates. Flak is expected to be heavy. Fighter escort at St. Catherine's Point at 0820. Take off at 0800. Good luck men."

As West arrived on the apron, the crews were unloading from the trucks which had pulled into the hardstand areas, and men were hurriedly loading their gear into the aircraft. Soon the noise of forty aircraft around the various squadron hardstands made the field come alive with activity. West's plane taxied to the runway and lined up with the others. The green signal flare arched over the runway and the planes began their take off runs to get airborne and head for enemy territory. The flight was uneventful until just before reaching the target when flak began bursting all around. Quinn saw Steere's plane get a flak hit in the left wing, and others were smoking from engine hits and structural damage.

Quinn's voice came over the intercom, "Daoust, have you picked up the target?"

"Yes, Sir, we're right on it ... Hold her steady ... Bombs away ... It looks good, Capt'n, let's get out of here."

West swung into a hard left and got into the formation heading back to base. The men breathed a little easier, but still looked for enemy fighter aircraft. They had nearly made it. Just another hour and they would be back at Rivenhall. The flight seemed long but after a while the Channel came into view, then later their familiar field was in sight. Quinn made the turns to put him into the pattern and bring his ship into the hardstands.

He could see from his left window Robbie holding onto their dog. Jiggs was jumping to try and get out to the plane but Robbie wasn't going to let him go until after the engines had stopped. When Quinn cut the engines and the crew began to unload, Jiggs bolted away from Robbie and scampered to the crewmen. They picked him up and laughed as they petted him, but Jiggs was looking for another one of the crew who had not come out yet. Finally Capt'n West came down and Jiggs ran over to him with that stubby tail wagging for all it was worth.

Quinn picked him up, "Hey, boy, have you been waiting for me? No, no, none of that face licking ... I know you've missed me ... Come on, let's go see if Robbie has any food over here for you."

The crew were in high spirits laughing at Jiggs' antics. Quinn was holding his hand up trying to get Jiggs to jump high.

West walked over with a broad grin on his face and told the crew, "I've got some good news for you men ... There were a few crews chosen for ten day 'flak-leave' to the U.S.O. hostels in England or Scotland, and you guys are one of the crews."

"Hot-dog! I can't believe it ... Can you Nat?"

The men were shouting and laughing, slapping each other on the back, the ground crew were smiling at all the hilarity. They could not go with the men on leave, but they were glad that their air crew were getting a break to rest for a while.

"Pick up your leave papers at Operations tonight and you are free for ten whole days ... Robbie, take care of Jiggs for us ... We'll see you guys when we get back."

"Okay, Capt'n."

"Captain, I meant to tell you sooner, but last week there were two young English boys who came onto the airfield and talked to us for a while. They had ridden their bikes from Silver End and came through the hedgerow to see our aircraft. They are real nice kids, and I think they are brothers. One is about 12 years old, the other is older. They didn't get in the way at all, but just stood back and watched us work on By-Golly. I enjoyed talking to them, but most of the men just kept working on the ship. We had some candy, and some things we gave them. They really seemed to appreciate it all. They had even brought an autograph book along and wanted us to sign it. I couldn't believe it, but it really made me feel good to think those boys wanted my autograph."

"Well, Robbie, maybe those kids just liked you. They may be our strongest tie with England because they will probably remember this war long after others have forgotten. If our histories are ever written, it will be by youngsters like these who see us as heroes. Did they tell you their names?"

"The oldest one was named Bruce and I believe he called his younger brother, Pete."

"Well, there's two good English names for you. I hope that I'm around the area when they come back again."

Quinn was there when the boys came back again and they still remembered him many years after the meeting. Forty years later Bruce Stait would write a history of Rivenhall and prominently mention the exploits of Captain West, his crew, and his aircraft, "By-Golly" in the history.

CHAPTER TEN

FLAK LEAVE IN SCOTLAND

That long awaited Friday in England was a special event in the lives of the "By-Golly" crew. They were starting their ten days "Flak Leave" as the men called it. Officially it was known as "R and R," or rest and recuperation, and it was earned after twenty combat missions. Actually the crew had flown twenty-one for their last mission was bombing the railroad bridge at Rouen, France, on the 31st of May.

They had gotten out of bed early that morning and headed to the mess hall together. Zola was ecstatic, "Nat, just think of it man, we are going to be free for ten days, I can't believe it."

"I can't either, but it sure is a good feeling to know we are going to be out of the war for a while just sight seeing in Scotland."

Zola questioned, "Are we all still set on going to Edinburgh?"

"Yes, Capt'n West sure wants to go. We've all gone to London on 3-day passes and seen everything, so it may be a good idea to see Scotland this time."

After chow they began to try and locate all the crew. They found Picklesimer and Budge on their way back to their quarters, and later Capt'n West and Fred came in. Everyone was excited. Zola was the first one to speak, "Capt'n, what are our plans for the next few days?"

"Well, it's really up to you guys as to where you want to spend your leave. For right now, I've arranged for a truck to take us to Chelmsford, and we can take the train to London. Whatever you want to do in London that night, whether a cinema or the U.S.O. Rainbow Club is your decision, then we get up early the next morning and catch the Royal Scotsman at Victoria Station. I really want to go to Scotland, but this is your vacation. A few of you may rather stay in London. If some of you single guys have girl friends there, then that will make a difference about where you want to take your leave. I see some smiles on a few faces. Okay, get your gear ready, and meet me at the motor pool at 10 o'clock sharp."

A couple of hours later, there was a crowd at the motor pool as Picklesimer and Daoust arrived. They soon discovered that some crews from the other squadrons of the 397th were also going on leave. Sgt. Picklesimer saw a friend of his in the group. "Hey, Neil, are you guys going on leave also?"

"Yeah, the whole crew is here," McGinnis answered. "Schubin and Mitchler are over by the trucks, and Captain Steere and Cramer are over talking to Captain West. Lt. Lipscomb is around here somewhere."

"Why don't you all ride with us in one of our trucks, and we can talk on the way."

The ride to Chelmsford was almost like a party with everyone laughing at each others tall tales. Nat was talking to McGinnis.

"Are you going to take your flak leave in London?"

Zola broke in, "Nat, you know McGinnis has got some girl friends in London ... Listen, Neil, what's the chance of you getting a few of us some dates in London?"

"Okay, I'll ask Katy. She and Ann have lots of friends there. We would all have a great time going to the cinema together."

Quinn and Cramer were sitting up front, smiling at the antics of the men trying to get their plans together for London. About that time the trucks pulled into the rail station at Chelmsford, and West called out to the men in the back, "All right men, all out for Chelmsford. We've got a train to catch in about twenty minutes."

The small railroad station got crowded quickly as all the men reached the area, but the hilarity continued with everyone talking and joking with each other. It wasn't long until someone yelled, "There she comes," and the crowd began to stream over to watch as the engine pulled beside the station.

As the train slowed, brakes began to screech, and the cars jolted to a stop. Zola hollered, "I get a seat by the window."

Nat answered, "Not unless you beat me to it ... Hey ... Look ... There's Captain Barnett's crew ... Donzello, come on in, boy, and get with us ... How about that? We are really going to have a party in London."

A sharp whistle blast, and the train began to slowly move away from the station. It was only about an hour to London, but the train made so many stops that sometimes the trip was stretched into nearly two hours.

The farmlands of Essex were beautiful to Quinn. It always reminded him of his home. They would pass small villages occasionally with narrow lanes and small farm houses dotting the landscape. Many times cattle could be seen grazing in some of the meadow areas beyond the houses and barns. It was pure contentment for him to just daydream and imagine that he was going home. Some of the larger towns like Chelmsford would have a modern look with shops, apartments, homes in the suburbs, and industrial areas scattered about, but mostly the County or Shire of Essex was farm country.

As time passed, the outskirts of London began to appear. From the small bungalows, the shops, and the flats of the suburbs to the larger industrial areas, one could tell with each change of building style that they were getting closer to the city. The train began slowly to pull into the confines of London. There were several stations called along the way, and finally Victoria Station was called and all the men started getting their baggage together. A few long whistle blasts and the passengers knew they were close. Then there was the hiss of steam, the screech of brakes, and finally a lurch and jolting of cars, and the men all began to pour out of the compartments.

Zola was full of questions. "Neil, how are we going to know if you can get dates for us this week?"

"I'll be staying at the Red Cross rooms, but we can get together later tonight at the Rainbow U.S.O. Club and I should know something by then. Will most of you be staying at hotels close to the Club?"

"Yes, I think so, but we still aren't sure about whether we will go to Scotland or stay in London for our flak leave. If you think that Katy might get us some dates, then Boy! that makes up my mind."

"Well, I'll tell you something tonight, and if things don't work out, then you can still make the train for Scotland tomorrow."

Captain West had left the group earlier to go and check in at the Regent Palace Hotel and get some sleep. He had planned to eat at the hotel's Pub, and later maybe write a few letters, then get up early the next morning to catch the train to Scotland. He smiled as he thought of the crew's excitement over London, and their wanting to do everything at once. He was a little older than most of them and more settled. Being happily married made him choose the quieter places. He was sure most of the crew would stay in London, but he still wanted to see Scotland with the historic city of Edinburgh and its castles of Holyrood, Edinburgh, and St. James Palaces. Sir Walter Scott had lived there and John Knox also. It was just a place that appealed to his interest. He had decided not to stay at the Red Cross Hostel so he had reserved a room at the Royal Caledonian Hotel. It was located in the center of town, thus he would be close to all the places of interest.

After dinner he decided to lay down and rest for a while. He could write his letters later. Then sleep began to take over and it was early morning before he awoke. And then, only because he had left a call at the desk to be awakened at 6 AM. It was fortunate they called him for he might not have awakened that soon, and been up in time to get dressed and make his train schedule.

He dressed and had breakfast in the Grill downstairs, then he made his way to the rail station. It wasn't long before the train's scheduled departure, so he made his way through the crowd. He turned downstairs to the tracks, looking for some of the crew, but not really expecting to see them.

The Royal Scotsman was a fine train and a fast one also. It made very few stops and was an express from London to Edinburgh. The compartments were well appointed, and it carried a dining car, so this was riding in luxury.

The trip would not be a long one for Edinburgh was only a little more than 300 miles from London. The route would take them through Sheffield and Leeds, then through the Cheviot Hills bordering Scotland and on into Edinburgh.

He found his seat and compartment, then placed his B-4 bag beside him. There were other Air Force men in his compartment who were also going to Scotland, and Quinn opened the conversation.

"Sergeant, what outfit are you from?"

"The 394th Bomb Group at Earles Colne, Sir."

"That's a good bunch. Is Colonel Hall keeping you straight?"

"Yes, Sir, he's doing a good job of that."

"My outfit is the 397th at Chelmsford, not far from your Field."

"Yes, Sir, we fly on lots of the same missions as the 397th."

The train gave a blast on the whistle and began to pull out of the station slowly. It started rolling faster as it moved through several stations, giving an authoritative scream of the whistle as it passed by each one.

England was beautiful this time of year, from the rolling hills and fertile farmlands of the Southeast all the way to the Midlands, it was green and scenic. You could hardly tell there was a war here, as the farms seemed so peaceful with cattle grazing on a thousand hills. This was the scenery that Quinn loved to see. After the war he wanted to go back to his home town, buy some land and raise cattle. The low green hills of his home in Northwest Mississippi were perfect for pasture land. His family had large tracts of land there that were used for farming and the dairy industry. There were Polled Hereford and Angus ranches all the way from Hernando to Jackson, Mississippi, as that area was a center for the cattle industry.

His son, Johnny, would soon be old enough to help with the work, and it gave Quinn a thrill to think of his farm's future name. He would call it, "J. Q. West, Jr., and Sons, Polled Hereford Ranch, Sardis, Mississippi." That sounded good to him as he leaned back and closed his eyes trying to visualize how his ranch would look. It was an American dream that Quinn never tired of dreaming.

The train raced through the countryside at a fast pace. It slowed through the towns and villages, but through the open country it fairly flew over the rails. The speed continued until they reached Sheffield, then the train began to slow as they made their way through the railyards and on to a stop at the station. Some passengers were boarding there possibly on their way to Scotland, but it wasn't long before the train began to pull out of the station. This was one of the few stops they would make, and at every crossing nearer the outskirts of Sheffield, they were picking up speed. The telegraph poles began to blur as she barreled along with the engineer hanging on the whistle cord as they rounded a bend or approached crossings.

The countryside was slipping by quickly, passing fields of wheat and other grain crops, open meadows, shining rivers, and small villages. The route took them through Leeds, the North York moors, and on to Middlesbrough. The next large town would be Newcastle-on-Tyne which was very near the coast and located on the River Tyne. The train crossed the narrow river and pulled into a picturesque station. There was an exchange of passengers, some getting off and others boarding for Scotland.

The route continued to follow the coast for miles through low lying foothills of the Cheviot Hills, the dividing line between Scotland and England. After crossing the River Tweed, the rest of the trip was through the Southern Uplands of Scotland. A rolling land of hills and green valleys dotted with farmlands and meadows. From here the land sloped downward to Edinburgh through rolling moors and occasional rocky cliffs. The train began to slow as the outskirts of Edinburgh appeared. Tidy suburbs and clean streets and crossings were passed just before getting into the city itself.

The train had slowed to enter the railyards at Edinburgh, as Quinn continued to look out the window, fascinated by the different appearance of the city. The train conductor called out, "Edinburgh Station in 5 minutes." The train pulled into the station with a smooth stop. No jerking or lurching. This was the queen of the rails showing her passengers that she was true royalty.

The crowd moved past the station slowly, and Quinn began to look for some booth or office for information about the area of Edinburgh. The Red Cross office looked like a good place to ask questions so he inquired, "Pardon me, Miss, could you tell me how to get to the Royal Caledonian Hotel?"

"Yes, Sir, Princes Street is just two blocks East. You can't miss it once you find the street. Here is a map of the city, and I'll circle the Caledonian for you. I hope you will visit our Red Cross Canteens in the city. There are three here and they will be glad to assist in helping you find the best sight seeing places."

"Thank you very much for being so helpful," he replied.

Quinn walked by the station and onto the street as he began to look at the city. This place was very different from London, a city of course, yet it had a kind of serene appearance with parks and wide streets. It didn't give the same look of crowding as did London, and it seemed to be completely untouched by the war. No bombed out buildings, no clanging fire engines, and no scaring or debris of the aftermath of bombing raids.

The Caledonian wasn't hard to find as several blocks later he saw it. The building was beautiful, just as he had hoped for; a large storied structure with comfortable rooms, and a Grill and Pub to have some delicious meals. As he checked in, the desk clerk told him, "Sir, I hope you enjoy your stay. The Grill is to your left, and the food is very good there. Please let us know if we can be of assistance to you."

"Thank you," Quinn replied. He hadn't been used to this type of courtesy in many months, but it was a welcome change of routine.

It was just the place he had wanted it to be. He could stay alone in his room and write letters if he wanted that, and yet it was close to the sight-seeing attractions if he chose to go walking. Already the city was attracting him. It was not a modern, high living society, but a quiet historical awareness of itself that made the city look timeless and beautiful. It was very different from the American cities, but it was just what he expected Edinburgh to look like.

His room was very nice with a freshly made bed, a writing desk and several chairs. The overall appearance was very appealing to him. It was a little early for supper so it was a good chance for him to lie down a few minutes, unwind, and close his eyes for a second. It turned out to be a long second for when he awoke it was nearly 9 o'clock PM. He couldn't believe he had slept that long. Now he was really hungry. He decided to walk to the U.S.O. Canteen on Princes Street, and see if they had some appetizing foods on the menu.

The canteen was filled with servicemen as he made his way through the crowd to the grill. He was getting ready to order when he heard a welcoming voice shout, "Captain West, where's the By-Golly crew?"

He turned around and recognized a gunner in his squadron, Sergeant Ray Snow. "Hello, Ray, I think the crew may be coming later. Are Stangle and Crummett with you?"

"No, Sir, most of the crew decided to take flak leave in London, but Sears and Skarles are here. We're at the Red Cross Flak House just outside of town. It's a beautiful two story home with lots of pretty Red Cross hostesses arranging baseball, tennis, and everything. Are you going to stay there?"

"No, I'm checked into the Royal Caledonian. I thought I could get more rest there."

"That's for sure. There are days I just want to do nothing but lay in the sack for a few hours, but it's game time every day at the hostel. It's supposed to help you forget about the flak, but it will take longer than ten days to make me forget about that stuff."

"Well, Ray, this is my first night here at the U.S.O. Tell me what foods on the menu taste like some good home cooking."

"I like their grilled cheese sandwich and a coke. I can think I'm home when I bite into that. They have hamburgers here too, but they don't taste like the ones back home."

"The grilled cheese sounds good to me."

"Captain, there's a whole bunch of the 598th men here tonight. 'Stretch' Gauker is over there talking to some friends ... Hey, Stretch, come here a second. You know Captain West don't you? Stretch flies with our Group."

"Sure I know Sergeant Gauker. You fly with Lt. Silverbach. We've flown together on a lot of missions. Are Taylor and Lee here this week?"

"No, Sir, they wanted to stay in London."

"My crew may be there too. I haven't seen any of them yet."

"Captain, we have a table over there. Won't you join us?"

"I appreciate the invitation, but I really need to catch up on some letter writing."

"Okay, I know what you mean. Captain, don't forget to see some of the sight-seeing places in town this week. You'll like the tour they give at Holyrood Palace and also the Edinburgh Castle. We'll probably run into you again if you eat supper at the Service Canteen on West Princes or the Red Cross Club on Regents Street. If we are not at either of these two places, we'll usually be at the Great Eastern Hotel club rooms after supper."

"Okay, I'll remember. See you later."

"So long, Captain."

After eating, Quinn walked back slowly to the hotel. He really did want to get a few letters off, but most of all he just wanted to hit the bed and catch up on some of the sleep he had missed the past few months staying up late, planning early morning bombing missions. Maybe he would just wait and write tomorrow.

He was getting sleepy, and tomorrow was another day. Possibly he would take a tour of the castles in Edinburgh in the morning.

The morning had dawned before he awakened, and as he pulled the curtains to look out he saw a lovely day beginning. What a perfect day to walk outside. He dressed and went down to the grill for breakfast. Then he headed down Princes and discovered the most unusual shops, bookstores and clothing establishments. Later he found the Sir Walter Scott monument and the Scottish National War Memorial. The Haig Memorial was near, so he went by there on his way to lunch at the Red Cross Club.

The Club was filled with servicemen when he got there, but none of the others were there that he knew so he ate quickly and left to visit Alexander Graham Bell's birthplace, the John Knox Home on Canongate, and the Robert Louis Stevenson Home at 14 Heriot Row. It had been a full day and a day of many historical sights, still he was becoming tired from all the walking and he turned down Princes Street and headed for the Caledonian.

As he reached his room and laid down across the bed he thought of how mentally refreshing the day had been. This was the kind of rest he needed. The strain of command responsibility had taken its toll, and he was beginning to feel this tiredness. The rest came just in time, and if the men didn't make it to Edinburgh, then maybe it was better this way. To be caught in a seven day sight-seeing tour would not be the type of rest he was looking for. It was the desire for rest from orders, from responsibility, and from the demands of military duties which he sought.

The following day brought a new chance to see some of the other attractions of the city. The morning was spent in a tour of Holyrood Palace and Chapel, Edinburgh Castle, and St. James Palace. He was learning how to get around in the city more and he was riding the double decker buses to implement his tours.

After lunch at the Caledonian Grill Room, he made his way to some of the Y.M.C.A. club rooms. There was one on St. Andrew Street, and another on Queen Street near St. Davids. Also the men had talked about a bridge spanning the Firth of Forth estuary that was unusual. When he reached the bridge he knew why the men had been interested in it. There were large stone supports rising from the water and a uniquely intricate steel geometric design spanning between each support. It was a beauty. As he gazed at the structure he wondered how many finely structured bridges he and his Group had destroyed in France. It all seemed such a shame to destroy the marvels of man's skill and ingenuity, but war demanded the price of total destruction of Germany's power to transport arms and material, and bridges were a vital link in their transportation system. The irony of it all was that it took years of struggle to construct these man made wonders, and only a few seconds of well placed bombing to destroy it all.

The days were passing swiftly and developed some wonderful memories of freedom from the cares of war, and images of so many beautiful sights. Several nights Quinn had gone to the cinema. There was the St. Andrew Square Cinema, New Victoria Cinema, and the News Cinema on Princes Street. Most of the movies were old, but the newsreels were interesting. It kept him informed on the fighting in Italy and the Japanese theater of war. The bombing of German occupied France from airbases in England were prominent news items. Everyone felt like an invasion of France was coming, but no one knew when or where it would take place. Quinn knew from his vantage point that bombing operations were intensifying with two and three missions each day bombing railheads, bridges, ammunition dumps, and supply depots from the French coast at Normandy all the way to Paris, so he was eagerly awaiting some news of the coming invasion.

The next morning Quinn awoke and heard the shouting of newsboys in the street selling papers with all the news about an invasion called "D-Day." It had happened June 6 and the "Scotsman" newspaper carried most of the details. Normandy was the invasion site of literally multiplied thousands of men, ships, and aircraft that had been gathered by the Allies for the coordinated attack. It gave Quinn a feeling of despair to be left out of such a grand maneuver. He was elated that the invasion had finally come, but he wanted to be there in the middle of it all. His training and skill had been a preparation for this battle and he was frustrated to think that he was not there to take part in the war. The news articles gave the full story. General Eisenhower had given the go-ahead for the attack at Normandy. West thought, what a place to be ... on rest leave while all the world was watching the action of the battle in France. He wanted very much to be there and see the great panorama of all that was happening across the Channel. He knew the By-Golly crew would be having a "fit" to get back to Rivenhall and be a part of the show, but it was not to be. Their leave would not be over until a few more days, and he knew it would gaul them to miss the action and they had worked so hard these past few months to help make it all happen.

The newspapers gave glowing accounts of the Marauders of his Bomb Group. They had led the attack by flying in low over the beaches and bombing just minutes before the assault troops came ashore. Quinn could nearly visualize the action. Col. Coiner was probably scheduling four missions a day and using every man he had to pulverize the German defended areas just ahead of the advancing American troops.

It was like a great shout of victory, and he was missing the battle. Maybe he could find the men at one of the service clubs, and see if any of them had been recalled to England. He hurried down Princes Street to the Red Cross Club and there he found them with a large group of servicemen celebrating the good news of D-Day. Everyone in Scotland seemed to be talking about it.

Quinn saw Sergeant Snow at the grill and he called out, "Hey, Ray, where are the others?"

"They're over by the juke-box celebrating with some Glenn Miller tunes. How about the D-Day news?"

"Yeah, it's great, but we're missing it all. That's the worst part of it."

"Well, there will be plenty to do when we get back. It's not over yet. I'll bet Coiner is keeping those guys in the air 24 hours a day. It must be bedlam for the ground crews trying to keep loading the bomb racks every time those flights come in empty. They probably load up and turn right back around and head to France again."

"Ray, you're making me feel a little better about it all. I was getting upset thinking I was missing it all."

"Yeah, me too, but we'll have our chance later. Here come Stretch and Art Kitrick now."

Stretch was smiling as he spoke, "Hi, Captain. What do you think about the news?"

"It's great isn't it?"

"Yeah, you know what bothers me most? It's that some other crews are out there flying our plane and they might not take care of her like we do."

"Oh me! Stretch, you're a card. I hadn't even thought of that worry, but now I'm feeling better after talking to you all. Come on and I'll buy you guys a coke."

The few days of remaining leave time seemed to pass quickly. Quinn packed his clothes on the last day and headed for the train station. He hadn't done all he had wanted to do while in Edinburgh, but he felt rested and ready to fly again, and that was important to him. He hadn't seen St. Giles Cathedral or St. Marys. He had seen the outside of the churches, but had not taken the tour. One thing he was very proud of was a photo of himself in the Royal Scots regalia. Ray Snow had told him about the photo shop on Princes that could dress you in Scottish uniform of kilts, sword, knee socks, bearskin hat, and all the trimmings of a genuine Scots Guard uniform, and then take a photo of you that looked like a real Scotsman.

The train was on time and the trip to London gave him a chance to lean toward the window and just enjoy the scenery as he daydreamed of the green hills of his homeland. The rural scene was not too different from his own farm. Maybe here a little less mechanized, but the farmers still planted, harvested, and raised livestock. The land was their livelihood, and it made little difference whether they raised rye, and we raised wheat. All farmers had a kinship. They loved the land they cultivated.

As the train moved through the countryside, sometimes a flock of sheep on a green stretch of grassland would bring back memories of the farm back home. Now and then a small community would appear, then later a more densely settled area until finally the definite confines of a large city were passing by his window. Buildings of every description merged together, then a slowing of the cars, and at last a voice of the conductor, "Victoria Station in ten minutes." The cars slowed to a snail's pace now, then a hissing of steam, and the screeching of brakes, and a final lurch forward and stop, let the passengers know they had arrived in London.

Quinn had slept nearly the whole way and it seemed as if the trip had taken only a moment. He straightened his tie as he walked to the station and asked for times of train departures to Rivenhall and Chelmsford. There was only one train in the evening so he bought his ticket and waited, looking over the London newspapers for any late news about the landings in France. Allied troops had pushed to twenty miles inland in some sectors, and the Marauders were concentrating on bombing German supply routes to cut off supplies of the retreating enemy.

He was deep in his concentration about the news when he heard a shout, "Captain West!" It was Zola, Natanek, and Picklesimer walking over to greet him.

"Captain, we sure wish you had been with us. Gosh, did we have a great time."

"It looks as if you had a good time with all those smiles and enthusiasm. What happened to your trip to Scotland?" Quinn asked smiling as they all began greeting each other with handshakes.

"Captain, we overslept and missed the train, then we thought if we waited until the next day we would be wasting some of our leave time, so we just said, what the heck, we'll check into a London hotel and take our leave here."

"Well, judging from the warm welcome, maybe it was best for us to take separate leaves." The men began to laugh after Quinn's comment.

"You won't believe it, Captain, but we really missed you."

"I'll have to admit I missed all the gang too. And D-Day, what about that? I felt like I was missing the whole war."

"Yeah, we did too, and now we're wondering what happened to By-Golly. She was probably flying every day while we were gone, and maybe even got ... No I'm not going to say it, but the first thing I want to do when I hit the barracks is find Robbie, and get a rundown on all that took place while we were gone."

"Did you see Daoust, Cramer, or Budge in London?"

"Yes, Sir, we ran into Daoust at the Regent Palace Hotel club room, but we didn't see the others."

"Well, Budge and the others may have gone back to the base early when they found out about D-Day. Zola, you and Nat tell me all about your stay in London."

"Captain, it was great. I couldn't begin to remember all the things we did, but we stayed at the Regent Palace Hotel on Piccadilly Circus, and spent lots of nights at the Rainbow Corner. The Stage Door Canteen was close by on Piccadilly West. We went there one night with some girls that Neil's girl friend, Katy, got us dates with. We went to several cinemas and a stage play at the Saville Theater on Shaftesbury Avenue. Nat, what else did we do that you remember?"

"We went on several tours of London during the day and saw most of the things we had seen on our two day passes, but it was still interesting. We went by the Victoria War Memorial, the British Admiralty, and Scotland Yard. I never get tired of seeing Buckingham Palace and the Kings Band playing, the Formal Guard Mount, the Old Guard and the Royal Scots passing in review. Hey, I forgot to tell you, we saw Capt'n Steere and Lt. Crummett at the Great Eastern Hotel Club Rooms on Liverpool Street. You saw Ray Snow and the rest of their crew in Scotland, didn't you, Captain?"

"Yeah, the whole bunch were there. Kitrick, Coyne, and Adams all asked about you guys. Did Sergeant McGinnis and his crew make the tours with all of you?"

"They did at first, but their crew only had 48 hour passes and they went back to Chelmsford on Monday."

"Okay, that's why Steere didn't go to Scotland."

"Captain, you haven't told us much about Scotland."

"Well, I had a good time there and visited some historic places like Sir Walter Scott's home, Alexander Graham Bell's home, and the John Knox home. There were also many historic castles there. Holyrood and Edinburgh Castle were beautiful. I just mainly had plenty of good meals at the Red Cross Canteen and I got plenty of rest at the hotel. Hey, isn't it time for our train to Chelmsford? Let's go ... We can talk on the train ride home."

The short trip to Chelmsford was like old home week with the men all trying to tell their stories first. Quinn told them more about his tour of Edinburgh, and they told him about all the U.S.O. shows and dances they had been to during their leave. They were all in high spirits and all anxious to get back to their outfit at Rivenhall. It was a three mile walk from Chelmsford to the base, but luckily they flagged down a G.I. truck and caught a ride in to Rivenhall.

The men reached the mess hall and started looking for Robbie. They spotted Sergeant Henry at a close table and Zola yelled, "Hey, Bill, have you seen Robbie lately?"

"No, but where have you guys been hiding out for the past week? You lucky dogs, you missed all the D-Day missions."

"Lucky? ... Not lucky, but just left out. We wanted to be flying when we read about the landings at Normandy."

"Well, I would have traded places with you. They just about worked us to death. I flew on three missions a day for nearly a week straight. I've never seen anything like it."

Robbie had been looking for the By-Golly crew and as he came through the mess hall searching for them he heard a group at the far table hollering for him.

"Hey, Robbie, come over here and tell us about our ship. Is she okay? What crews have been flying her while we were gone? Did she catch any flak or get damaged on the bombing missions?

Come on now and tell us the truth."

It was hard for Robbie to begin answering questions with all the handshakes and the men clapping him on the back with unrestrained welcome. "Wait a minute you guys ... Wait a minute. She's okay. She flew six times while you were gone. Lt. Barnett and Capt. Williams flew her one time each, then Capt. Stangle flew her two times and Lt. Taylor also piloted her twice, and one of his flights was D-Day, June 6."

"How about that? She was in the fighting on D-Day. I'm feeling better. At least there was one of us out there on the big day."

"She caught a little flak on a couple of missions, but nothing serious. You guys are on the loading list for a mission tomorrow, and By-Golly is ready to go. Hey, tell me all about your flak leave? Did all of you go to Scotland with Capt'n West?"

"No, we missed a train and all of us stayed in London. We had a tremendous time though. Plenty of U.S.O. dances and some real good bands at the Rainbow Corner dances. It was great, really great, wait until we have time to tell you all about it."

Robbie replied, "I'll be having some leave time in a few weeks, so don't forget to tell me all about the places you went."

The men left the mess hall laughing and swapping stories of all the experiences they had while on leave. Soon it would be time to get back to the grind of missions and combat, but they were rested and in a way, were anxious to keep building their missions to fifty and return to the U.S.A. permanently. They had twenty-one missions now and only twenty-nine more to sweat out. It was reasonable to believe they would be home for Christmas. This was the hope that kept them going; the dream of home, and being with those who loved them. The remembrance of sweethearts, parents, wives, little brothers, and sisters. These were the precious faces they longed to see again.

If you asked them why they fought so hard, they would probably answer that they wanted to hurry and win a war, but behind it all was the face of their loved ones they wanted to see and embrace again.

A flight of 597th Squadron Marauders over Coggeshall, England, on their way back home to Rivenhall Air Base after a bombing mission.

CHAPTER ELEVEN

COMBAT MISSIONS, ST. LO TO NANTES

The briefing the next morning was set for 0700, but conversation was animated in the mess hall long before that as the men laughed and caught up on all the things which had happened while they were away on leave.

Zola asked some of the air crewmen, "What did you guys do while we were gone?"

"Well, we flew ten missions, and two of them were on "D-Day" at "Utah" beach and at Trouville ... Boy, you should have seen the ships in the Channel on June 6 ... It was amazing. We flew missions to Le Harve several times, and St. Lo and Le Mans. Did you know they canceled all passes the day after you all left? You just barely made it."

"We read about the landings in the papers and everyone of us felt like we were missing the whole war. Robbie, how did Jiggs get along without us?"

"He's doing fine, but I think he misses you guys 'cause some days he just seemed lifeless ... I'd like to be around to see what he does when you all show up."

"Did our aircraft really come through all right?"

"Yeah, she caught a little flak on a couple of missions, but no major damage ... Boy, they caught some flak on the Le Havre mission ... There were nineteen planes damaged, and twelve were damaged on the Rennes Mission ... They have been busy on the ground trying to keep the planes in service."

"We're ready to get back to flying ... It's strange, after we had been gone for a week we started thinking about our ground crews and some of our other buddies and we were ready to get back to the base."

"Yeah, we missed you guys too."

The briefing got started on time as Col. Coiner was leading the first part of the session."Our target is a rail junction at St. Lo and a very important one. The Germans are bringing in supplies through here to supply the troops fighting our men in Normandy. There have been some good gains by our ground troops in this area, but they need our help to knock out supply lines and supply dumps so the enemy won't have the material necessary to wage war. Take a look at this line ... It is our bomb line and it moves forward every day. Our troops are behind it and the Germans are in front of it, so never pick any casual targets behind that line.

The weather should be good over target. Bomb at 8,000 feet ... Fighter escort at 0820 ... Take all notations for your headings to target and return headings. Col. Allen and Dempster will lead this one. Take off at 0800."

The men were anxious to go ... They had seen the landings in Normandy, and they could tell the tide of war was moving in their direction. It made them more eager to finish the war and get home.

The trucks were loaded and began to pull off to the hardstands. Quinn's crew couldn't wait to see Jiggs and their ground crew. When they pulled up there he was and he didn't recognize them until they jumped out of the truck and began petting him. Jiggs got more excited by the minute as he began to know his friends had come back. When he saw Quinn they could hardly hold him ... He was squirming to get down and when they put him down he made a bee-line to Quinn and barked loudly.

"Okay, boy, come on," Quinn played rough with him. "You didn't think I'd come back did you? All right, Jiggs, calm down, big boy ... Go over and get Robbie to feed you ... We've got to go on a mission."

"Capt'n, sometimes I think you'll have to take this dog with you when you take off. I can barely hold him when your plane starts away from the hardstands."

"No, we can't ever take him, Robbie, it would kill me if I ever lost him on a mission ... I don't think he would understand the engine noise on a flight, and there's too many dangers in the air. He could fall out of the bomb bays or waist gun openings. I feel better when I know he's safe on the ground ... Is the ship in good condition?"

"Yes, Sir, we've kept her in good shape."

"Okay, let me get up to the flight deck and we'll start 'em going and blow the carbon out."

Quinn looked out his side window and slid it open. "All clear, Robbie?"

"Yes, Sir."

Quinn had already brought the inertia starter up to proper RPM. He clicked the mesh switch left and watched as the big four blade prop began to turn slowly, then a few cylinders began to fire and quickly a few coughs of white exhaust and she started hitting solid on all cylinders. He released the brakes after starting the number two engine and began to ease out of the hardstand area.

Once they were airborne, the mission went well until they began getting hit by flak over the target. Quinn's ship made it through the barrage okay, but some others got damaged. After the mission, the count was eight aircraft damaged, but the bombing was classed as excellent, so the results were worth the damage.

After that, there were several missions in quick succession. On June 14 they bombed a railroad bridge at Chartres with fair results and fifteen damaged aircraft. Then on the 15th they went back to the same target with poor to fair results and twelve aircraft damaged.

On the 18th of June a bombing mission was scheduled for a marshalling yard at Mezidon. The planes went in at 9,000 feet and the crew couldn't see a thing because of 10/10 weather, so the mission was recalled. They went out again on the evening of the same day to bomb a V-1 site near Bachimont, and the weather was so bad they set it up as a pathfinder mission. The results were not obtained as cloud cover prevented strike photos from being taken.

The pathfinder missions were flown by using a very sophisticated radar system in the lead aircraft. They could bomb through solid cloud cover without ever seeing the target. The equipment in the plane was called a "Gee-Box" and it was operated by a radioman trained in the use of this specialized equipment. When the lead aircraft opened its bomb bay doors over the target the other planes did the same. Then as they saw the first bomb come from the pathfinder plane, they all dropped their bombs at the same time. It gave a good blanket or carpet effect that covered the entire target area.

At this time on most of the missions which Captain West flew, he had been selected as deputy lead, and his aircraft had the "Gee" equipment installed in it. There were two Gee Operators who consistently flew with him, Sgt. William Henry or either Sgt. Lloyd Webb.

To be able to bomb through the clouds seemed like the ultimate solution to aircraft damage and losses, but the Germans also had improved their equipment. Consequently, they were aiming the antiaircraft guns by radar, and the bombers' losses to flak were just as high or higher than ever.

On the 22nd of June a pathfinder mission to bomb a defended area near Nouainville was scheduled. The mission was successful but nineteen aircraft were damaged by enemy flak. Another pathfinder mission was scheduled the next day for a V-1 bomb site near Lambus, France, in the Pas de Calais area. The mission was successful with nearly all aircraft dropping on target and no damage.

The next day, the 24th of June, the "By-Golly" crew were not scheduled and they looked forward to some restful baseball and basketball games and maybe a night at the Aero Club or the movies. But those crews which had been scheduled were expecting a very tough mission.

The mission's target was the railroad bridge at Maisons Laffitte, one of the worst flak areas around Paris, and the men had learned by experience to dread this location. Captain Steere's crew had been scheduled for this mission, and the "By-Golly" crew were sweating out the mission for them. They were close friends with Steere's crew, and they waited for McGinnis, Schubin, and Mitchler to return so they could ask them about the mission. It was a long wait and an agonizing one, but when the ships finally began to come in, Zola and Nat could see they were in trouble. Stragglers were coming in from several directions and red emergency flares were everywhere. One plane hit and slid off the runway; another plane turned completely around and ran into trees at the end of the runway; another jumped a ditch on the side of the runway and was stuck there. The other aircraft came in with extensive damage to engines, wings, fuselage, and stabilizers. The emergency crews were trying to do their best to help wounded airmen and clear runways for landing aircraft. It was a scene to bring tears to the eyes of strong men. The 397th was suffering.

It took quite some time before the crewmen began coming from the debriefing rooms. Nat saw McGinnis and asked him what had happened on the mission.

"Nat, it was a mess all the way. We started getting flak about fifteen miles from the target. We were in the second box and we watched the first box lose four aircraft right off. We learned at the debriefing they were Gatewood, Knox, and Neill, from the 597th, and Captain Powers from the 599th. We had to go through the same flak and I thought we would never make it to the target. I looked around and every ship had an engine throwing smoke except ours. A piece of plexiglas fell on my head. That was a close one. When we finally got to the target there were only three aircraft. We salvoed our bombs in the area, and of the three ships left, ours was the only one whose engines weren't smoking. Carlson was flying lead and he feathered his right prop and fell out of formation. We swung to the right over Paris. We were supposed to go left, but there was too much flak over there. We were by ourselves, so Johnson gave us a heading and we went right back through the flak. The first box sunk the bridge. It was a good thing. I sure don't want to go back soon. Out of nowhere B-26's joined in formation with us. Our right govenor went out so we flew in fixed pitch, then fell out of formation near London. Our gears came down okay then we lost all hydraulic pressure. We needed flaps so I took off for the bomb bay to crank down flaps and got them down just in time. Steere landed, hit the air-bottle, and slid off the runway. Carlson made it to the beachhead. I was glad to hear it. Every plane was hit except one out of thirty-six ships. Boy, what a life."

"Neil, we saw your plane come in when it slid off the runway, and saw all those other damaged aircraft, also. It was awful. What was the final tally from S-2?"

"Well, as it stands right now, there were four ships lost over enemy territory, one man killed, three men wounded, twenty-five men missing in action, and thirty-three aircraft damaged. Boy, we can't stand many more missions like that."

The 397th didn't fly again for five days. It was impossible to repair all aircraft quickly after such battle damage, but on the 30th of June, West's crew were scheduled to fly again. The target was a bridge near Thury Harcourt, but 10/10 weather over the target recalled the mission and no bombs were dropped. That same afternoon they went on another mission using a pathfinder and bombed a road junction at Conde Sur Noireau.

Again the 397th didn't schedule a mission for five days, and the mission on July 6 to the Dol To Rennes area for the bombing of a railroad line was another costly mission with damage to eighteen aircraft.

West's crew was scheduled again on July 7 to bomb some motor transports near the Laval area, but weather caused a recall. It turned out to be a costly mission as flak shot down one aircraft over enemy territory and damaged twenty-four other planes. One man was wounded and eight men missing in action. They were scheduled again on the next day to bomb a railroad bridge at Saumur and the results were good; however, seven planes were damaged and one man wounded. They were sent back to the same target only to find weather so bad that the mission was recalled after seven planes were damaged by flak.

They were scheduled again on July 11 to bomb a fuel dump at Chateau De Tertu. The bombing was good and no damage to any of their aircraft by flak. Then the "By-Golly" crew got a four day run for some much needed rest. Forty-eight hour passes were there for the ones wanting to go to Chelmsford or Braintree to the shops and pubs of the towns. Others decided to stay at the air field and catch up on sack time.

Quinn's crew was scheduled again on July 16 to bomb a railroad bridge at Nantes. This would be the 70th mission for the 397th, and the 35th mission for the "By-Golly" crew. It was to be a very difficult mission for them. There had already been one mission that morning, and this mission was in late evening at 1900.

Quinn had misgivings about the evening missions. He just didn't like them. The mornings seemed right, but the evening bombings didn't suit him. Why, he really couldn't tell. On this mission, West's plane was hit badly by flak over the target area. His right engine was hit and a large flak hole in his left wing tank caused him to pull out of formation and look for an American airfield in Normandy. He belly landed his aircraft on a fighter base just built after the "D" day landings in Normandy and saved his crew from a bail out situation, but they had to spend several days in France before their outfit could arrange for transportation. They had lost their plane for the stricken "By-Golly" had burned completely after all the crew had escaped. Now the plane which had carried them so gallantly into battle was gone. It was a great loss to all of them.

Quinn would later be awarded the Distinguished Flying Cross for his skill in handling his damaged aircraft on this mission.

The crew was having a difficult time trying to get transportation back to their field in England. The Fighter base in Normandy didn't have transport aircraft to get them back, and their Group at Rivenhall were much too busy to arrange for transportation, so they just waited.

Finally after several days a plane arrived from Rivenhall to take them back to their home field. It was a happy occasion to be again in England. All the men were asking about the circumstances involved in the belly landing and crash while Quinn's crew were busy asking questions about how the 397th had faired during their absence.

"Nat, what happened to "By-Golly?"

"I hate to say it, Robbie, but she burned almost completely. The crash trucks got there soon enough to save some of it, but the engines and cockpit area were mostly burned out. There wasn't much left to salvage except from the wings back to the tail section. They took some photos of us standing around the plane. Boy, it was a mess with foam sprayed everywhere."

"Zola, you guys are going to miss your ship."

"You bet we will. She was a good one. I knew we were done for when flak hit that right engine so hard. Hydraulic fluid was everywhere and then we discovered that fuel was pouring out of our right tank. It was terrible. We started pitching everything out and someone grabbed Cramer's chute and Cramer yelled, "Wait a minute, that's my chute." It seems funny now, but at the time I think we were too busy to laugh."

"It's funny now to hear you tell it, but I'm glad Cramer didn't lose his chute."

"Where is Budge and Captain West?"

"They went to Headquarters. I think Col. Coiner wanted to talk to them."

"I wonder if Coiner is upset over our crash."

"No, nothing like that. I think they have already put in a commendation for West for the way he handled the aircraft?"

"I'll say one thing, he deserves it. You should have seen him belly that Marauder in. We didn't have any hydraulics. The plane was at stall speed the whole time and just hanging in the air by a thread. When she hit, you could hear screeching and bottom panels shearing off, but it was a controlled crash, because West kept her straight down the runway and didn't let the wings touch and cartwheel us into a real bad crash. If the plane hadn't caught fire from the right engine and burned so badly we could have salvaged ninety percent of it. I've seen them put planes back in service that were a lot worse than ours. Well anyway, she's gone now and we'll miss her."

The headquarters building was a large quonset type of structure at the entrance to the air base. Quinn and Budge walked to the front desk and asked an orderly if Colonel Coiner would see them. The orderly nodded yes, and they followed him to a row of offices, one marked Col. Coiner. The orderly knocked and a voice boomed, "Come in."

West and Budge walked through and saluted as they saw the Colonel.

"Sit down gentlemen. I heard you had a bad time over Nantes."

"Yes, Sir, we had a good bomb drop, but we caught it after we turned back to the field. I'm just sorry we couldn't have brought her back to Rivenhall."

"Well, I am too, but it couldn't have been done any other way according to my reports. I'm proud of the way you belly landed in and saved your crew... You're a good pilot, West, one of the best I've got."

"Thank you, Sir."

"There are two things I want to tell you ... One is that I have written a commendation for your skillful handling of a badly damaged aircraft, and the other is that you and your crew will be issued 48 hour passes to London... Have you ever heard of a radio program called, "American Eagle in Britain?"

"Yes, Sir."

"Okay, they want your crew to appear on the program Saturday and relate your story of the belly landing in Normandy."

"Thank you, Sir, we appreciate it."

"All right ... Budge take care of this guy, he's too good a pilot to lose."

"Yes, Sir."

Budge and West walked together to the officer's quarters. They could hardly believe what had happened.

"Budge, those guys will be so excited about being on the radio they won't know what to do."

"Yeah, we'll all be speechless except Zola. He'll have to do the talking."

"That's right." Budge and West were both laughing.

It was time for noon chow and Quinn thought he would see them at the mess hall. It wasn't long before they came in and West gave them the good news.

"Boy, I must be dreaming," Zola shouted, "the whole crew is going?"

"That's right, our whole crew."

"When do we leave?"

"Tomorrow, because the program is going to be taped on Friday."

"Captain, I'm going to pack tonight, and probably stay up half the night talking to Pic and Nat about what we'll do in London."

"Don't stay up too late, you might fall asleep at the broadcast."

They all laughed, as several others of the crew gathered around to hear about the plans from Captain West.

"Look, we're going to catch our train at Braintree tomorrow because the Chelmsford train schedule doesn't get us into London early enough. Nat, I want you to talk to your buddies at the motor pool and arrange for two trucks to the station. If they have some large trucks that will be even better, but get what you can, and if you have any trouble, let me know, so I can clear it through Col. Coiner. There are some other crews who have three day passes this weekend, so they may want to ride with us and get the early train. But just be sure we have plenty of transportation. Put on your best Class "A" uniforms, and let's show them what a good-looking group these 397th men really are. All right, turn into bed early, and I'll see all you guys tomorrow morning for chow. Zola, practice your lines. I know they will be talking to you." Everyone laughed, as they began leaving.

CHAPTER TWELVE

LONDON AND THE AMERICAN EAGLE

The mess hall was beginning to fill with the usual noisy conversations and laughter of the morning chow call. Donzello was especially exuberant as he yelled at Zola, "Hey, buddy, how many on your crew got three day passes?"

"We all did, son, even Capt'n West is going and that's something. I believe Capt'n Stangles' and Ryherds crew are going too."

"Yeah, I heard about them ... We are really going to have a party, boy, London will never be the same after this one."

"Has anybody seen McGinnis? He is supposed to be getting his girl friend to look for a few dates for us."

"We'll probably run into him at Braintree when we catch the train. Are you guys sure we can catch a truck to get us to the station? I sure don't want to miss that train and have to wait until tomorrow."

"Don't sweat it Donzello, I've got it all arranged. Some of my buddies in the motor pool are going to Braintree to pick up supplies so we are going to meet them at the main gate at 8:30."

"Well, let's finish and get out of here ... We don't have much time."

The day had turned out beautiful and that was a relief as many days this time of year were overcast, cold or rainy. The group had made contact with their trucks and the ride to the station seemed to have a party atmosphere. Soon they were pulling up to the rail station and the trucks came to a jolting stop. Zola yelled, "All right you guys, pile out. Let's get our tickets and listen for that train whistle. There's McGinnis ... Go grab him and find out what the score is."

Nat and the others in the group were laughing, "Settle down, boy, you're going to blow a gasket in a minute. Hey, here comes the train ... She's blowing plenty of smoke ... Must be a heavy load today."

The train was running a little slower than usual but no one complained. It wasn't that far to London, and fast or slow, they would be getting there in a few hours. Braintree was a main station and many times the engine would take on water or coal, but today it by-passed the water tower and began to unload some twenty passengers or more for town and prepared to load the mad rush of servicemen when the conductor called for them.

Zola, Nat, and Robbie were in the middle of the crowd as they saw Ray Snow and Skarles boarding the main coach. "Hey, Ray, save us a seat, boy, get me one by the window."

"No way, pal, it's too crowded ... You're going to have to stand up."

It was crowded, but everyone found a seat. Many of the local passengers looked at the servicemen in disbelief. They seldom saw such an exuberant and rowdy group of young men. It seemed the Americans were always noisy and hardly ever restrained or conservative, but the locals accepted them ... They were a good bunch of boys, but just a little too boisterous for the English taste. The airmen looked at the local passengers also. That was half the fun of going to London to see the sights and the people, especially the pretty girls. The men were respectful to the English and probably a little fascinated by their ways and manner of speaking since most of the men realized that only a few generations separated them from these people, their English cousins.

The engine puffed leisurely while standing still with brakes locked and smoke billowing out of the stack, waiting for the conductor's signal for "All Aboard." Then she gave a sharp whistle blast and the hiss of brakes releasing as a forward jerking motion signaled that the engine was pulling slowly away from the station. This was a far cry from the swift wings of the B-26, but the men loved it. They allowed the swaying coaches and the hypnotizing clack, clack, of the rails to take them away from the pressures of war ... They welcomed it. Let the smell of coal smoke and steam take their minds back to their boyhood days in America. It was a feeling of contentment.

The conversations tapered off as they began to look at the scenes of the surrounding countryside. Cattle were grazing on the green hills of Essexshire. Sometimes there would be a farmer plowing, sometimes a farm girl carrying water or walking to a neighbor's house, and sometimes just children playing. It was the medicine they needed to forget a world at war.

"Hey Skarles, look at those kids down there waving at the train. Did you ever do that when you were a kid?"

"Sure, all the time ... I used to wave at trains."

"Yeah, me too ... Sometimes the engineer or fireman would wave back and that was great. I guess I wanted to be an engineer when I was a little kid. Later I built model airplanes and wanted to be a pilot ... But, that's how it goes."

"Hey, we're slowing down ... There must be a town ahead."

"Zola, it doesn't have to be a town ... This train stops for anything."

"Aw, don't give me that stuff. I've heard Schubin telling those tall tales about trains stopping for people waving a handkerchief ... I don't believe it."

"Okay, okay, have it your way, son, but we ain't got to London yet."

The guys started laughing at Zola and began punching on him in a kidding way, but he liked the good natured kidding. It was all part of the fun.

The men had begun to settle down just before pulling into London at Victoria Station. Some were dozing by the windows as the sun's rays warmed them while others were still talking over the constant clack of the rails and the occasional shrill whistle blast as they approached each small rail crossing and village. Nat was watching some children play in the isle of the coach when two very familiar figures came through the open door.

"Hey, here comes Capt'n West and Lt. Budge." Those that were dozing looked up and there were West and Budge smiling from ear to ear.

Zola said, "I thought both of you had missed the train."

Quinn answered, "No, we made it but we decided to get in the second coach so we could make some plans for the squadron. What have you guys been doing?"

"Oh, just talking about all the fun we are going to have in London. Will you and Budge make the rounds with us?"

"Well, we plan on eating a meal with all of you at the Regent Palace Hotel Restaurant, but I don't know about making the rounds. I'm an old married man," Quinn laughed, "and you guys will have more fun without me tagging along as a chaperon. Some of the officers are staying at the Regent. You know where it is on Piccadilly Circus. The restaurant has some great food ... What time do you think would be best? Will 8 o'clock tonight be okay for you guys? Budge and I are picking up the checks."

"Sounds good to us, about 8 o'clock and we'll be there. Most of us are going to try and get rooms at the Royal Hotel. All the men in the 397th hang out there in the Pub."

"Well, maybe one night we will get by there and see how you are all doing. It's at Woburn Place isn't it?"

"Yeah, Russell Square."

"Listen, when you come to the Regent tonight be sure you look for the restaurant because they have a Pub and a Grill Room, so don't get in the wrong place."

"Okay, don't forget 8 o'clock ... We'll see you men later."

The scenery outside the train windows was beginning to look like a city. The train was slowing and had already passed through suburbs miles back. Now the tracks were branching into rail yards as factories and storage buildings were beginning to predominate. Most of these were blackened from years of coal dust, smoke, and soot. A few newer buildings were interspersed and some bombed-out buildings were seen among the others. As the train slowed, more shattered structures were seen when Nat commented, "Man, look at those building shells. It looks like some Buzz-bombs zeroed in on this section."

"Yeah, maybe a few bombs started a fire and burned a whole section down."

Zola answered, "Man, that is terrible. I hope London hasn't gotten this much damage. Ayers and Rodi were here a few weeks ago and they said there was lots of damage in some parts of the city."

"You know, it looks like there wouldn't be a single V-1 left after all those bombs we dropped on the "Rocket" sites."

"Yeah, but those Germans are smart. They just move their launching ramps to a wooded area or some French farm houses and we can never spot them."

"Hey, I just thought of something ... Look at us, we are trying to get out of the war and we'll be dodging flying bombs the whole time we are in London."

"Aw, they say that people don't get hit much by them. It's just buildings that get destroyed."

"Yeah, but what if I'm in the building?"

"Okay, here's the deal ... We'll go to the theaters, pubs, and U.S.O. dances and you stand outside and wait for us, then you'll never get hit by a Buzz-bomb."

The crowd burst out laughing as Pic answered, "Okay, wise guy, I get the message. But you know McGinnis nearly got hit by one of those Doodle-Bugs when one blew up a building that he was standing near."

"Yeah, I heard about it ... He said the concussion made him feel funny for a while."

The train was slowing to a snail's pace and the tracks were widening into a large rail yard. After a few whistle blasts there were several jerking stops and the conductor came through their car and announced, "Victoria Station in two minutes."

After a while they could see the long platforms with baggage carts and freight carriers pulled along side waiting to unload the coaches. It would be a short walk to the Station, and then London.

The men were in high spirits as Zola led the conversation, "Man, I can hardly wait ... We are going to have a great time together."

"What are you guys going to do this afternoon, some sight-seeing?"

"Yeah, we might find a room first then head for some of the main things to see after we get our broadcast over with."

It wasn't easy to find the British Broadcasting Corporation. They had the address, but the area was not one with which they were familiar. However, after asking directions several times they found it. A receptionist escorted them to the sound-room, and they discovered that several other crews were there to be interviewed along with them. It gave them a chance to see how the other crews were being interviewed.

After a while the announcer called their crew and began his interview.

"Captain West, tell us what town you are from in America."

"Sardis, Mississippi."

"Would you spell that for us?"

"S..A..R..D..I..S."

"You and your crew fly the B-26 Marauders of the Ninth Air Force, do you not?"

"Yes, Sir, the best ship in the E.T.O."

"Your crew seem to share your opinion with so much cheering. Would you tell us something about your Unit?"

"Our Group is the 397th Bomb Group stationed in Essex near Chelmsford."

"Now, tell our listeners just what happened to your aircraft during that belly landing in Normandy."

"We had just completed our bomb run when flak hit us in our right engine and right wing tank. There were other hits that cut hydraulic lines and electrical lines, so our instrument panel was inoperative, and the wheels couldn't be brought down. We discovered an American Air Base near the beaches at Normandy, and we decided to try and belly land on the side of the runway."

"Sergeant Zola, tell us what you thought when your plane was belly landing."

"I was hoping John Q. knew what he was doing up front."

The crowd in the sound room were laughing at Zola's remark.

"Oh, does Captain West let you call him that?"

"Only when he can't overhear me."

The crowd started laughing again.

"Captain West, how many missions have you flown?"

"This one was our 35th."

"Congratulations, and I know you and your crew will have many more successful missions. I am curious about how your aircraft got the name "By-Golly." Sergeant Zola, could you tell us?"

"Well, Capt'n West doesn't use swear words, so we picked the strongest word he ever uses, and that was "By Golly." Also it happens to be the expression he uses most of the time when he sees or hears something amazing."

"Captain West, thank you and your crew for being with us today. Good luck to all of you men on your future missions."

After the interview the crew was told that their families would be notified back home when the program would be broadcast to America and that their families would also get a record of the interview. The men were elated that their parents would get to hear about the exploits of their crew, and they were proud of the honor of being on the program. As they walked out, Zola began asking where they would be going to next. Capt'n West and the officers headed to the hotel, as Nat reminded Zola what Neil had told him that morning.

"McGinnis said to be sure and visit Hyde Park ... He said the Marble Arch and the Lover's Bridge were beautiful with all the floral gardens around ... Then we're not far from Buckingham Palace and we can check in the area and find out what time the King's Band might perform ... I'd like to be there at the time for the formal guard mount with the old guard passing in review or possibly see the Royal Scots in review. It's fantastic, man, and really a precision military ceremony. But, whatever you guys do, just remember our dinner with Capt'n West at 8 o'clock at the Regent."

The time passed swiftly. There were just not enough hours in the day to see everything, but the men were happy to be away from the duties of the air base. They walked nearly everywhere, as most of the main sights were reasonably close. They could have ridden the doubledeckers or any number of other forms of transportation, but they chose to walk ... It was simpler that way.

By eight o'clock that evening they were near the Regent Hotel and getting hungry. Nat and Zola were walking up the front steps of the hotel entrance when they saw West and Budge walking toward them."Hey, I thought we were going to have trouble finding that restaurant and here you guys are at the front lobby."

"We thought you might be coming about this time so we came to the front to look for you. Come on, the restaurant is back this way. Are you all hungry?"

"Yes, Sir, we can't wait." They walked through the lobby and entered a large restaurant.

"Well, have a seat and let's take a look at the menu. Their steaks are especially good, but order anything you like." The meal was one of the best they had eaten in months with steak and all the trimmings.

Zola was keeping the conversation moving. "Capt'n, why don't you and Budge come with us to the Royal Hotel Pub tonight? All the guys are going to be there."

"Well, let me take a rain check on it tonight. I want to write some letters."

"Aw, Captain, write those letters when you get back to Rivenhall. See some of London while you are here."

"Oh, I'll do plenty of sight seeing this weekend, but it's tough for me to write letters at the air base because there are always so many other things to think of. Here it will be quiet in my room and I can just sit back and daydream about home and write all the things I want to say without interruptions."

"Yeah, I see what you mean ... I don't write home enough either, but I'm the first one in line at mail call and always disappointed when I don't get a letter.

"Zola, I'll tell you what ... I'll try to make it to the Pub with you guys tomorrow, or would you rather all of us went to the Rainbow Corner U.S.O.?"

"Yes, Sir, that sounds better."

"How about this time tomorrow?"

"Okay ... The meal was great Captain ... We sure appreciate you and Lt. Budge thinking about us."

"It's been our pleasure, Zola ... We don't get a chance to sit down and enjoy a good meal together very often. We need a few good memories to take home with us after the war. Hey, by the way, I have a surprise for you men if you haven't been to the Rainbow Corner in the last few weeks."

"All right, Capt'n, see you tomorrow."

Nat and Zola walked away wondering what kind of surprise the Captain was talking about.

Quinn turned to Budge and said, "Bill, they are a good bunch of men, aren't they? Sometimes I think I holler at them too much about mistakes they make."

"John, you don't really holler at any of us, but when they catch some flak from you they know they deserve it."

"Well, I hope they'll remember some of the good times."

"Sure they will ... We all will. When we get back to the States after this war we'll get together and throw some parties that will be out of sight ... The best is yet to come."

"Yeah, it's going to be great, Bill ... I guess I'll turn in for the night. You might want to head out with some of the other men, but I think I'll hit the sack."

"Okay, John, good night. I'll see you for sure at the Rainbow Corner."

"Sure thing, Bill ... Goodnight."

Quinn turned and walked up to his room. It was still early enough to write some letters. The room was quiet and he was in the mood to write to Ruby. He always thought about her, but today especially he had her on his mind as he toured London wishing that she could be there with him to see the beautiful sights. What a pleasure it would be to see her and his son, Johnny. He could close his eyes and see Johnny's big smile and hear his gleeful squeals as he would run to him and call his name loudly. He had some pictures of him in his wallet, but Johnny had probably grown a lot since they were taken. Maybe he was saying lots of new words now. Ruby's last letter had been a long one and she had written how Johnny was growing bigger and saying new words each day. Quinn began to drift into sleep as he closed his eyes and thought of home.

The day was bright with sunshine as he awoke and discovered it was nearly 10 o'clock. He hadn't planned to sleep that long, but he still had plenty of time to see the things that he wanted to see. He wanted to go again to Westminster Bridge and the surrounding area to see Westminster Abbey and Big Ben. It always gave him a thrill to hear it strike the hours. Scotland Yard was close and the buildings of Parliament. The Thames was still his special scene of beauty and he spent extra time there. This trip also he wanted to see the War Memorial in front of the British War Office and the Victoria War Memorial. There was so much to see. Maybe a few sights could be saved until his next trip to London, whenever that would be. The time passed quickly and even though he had skipped lunch to see a few extra things, it was late afternoon before he had realized it. Now, it was nearly time to head to the Rainbow Corner and meet Nat and Zola. Maybe there would be some of the other men and officers of the 397th there. It was a gathering place for all servicemen; English, American, and others, but it seemed that the Americans mostly liked the Rainbow Corner best. He knew that Stangle and Steere were in town and maybe Taylor and Lee. He walked faster down Regent Street and as he turned the corner he could tell the Club would probably be full. There was a crowd as he walked in and began to look for familiar faces. He spotted Bill Henry, one of his radio crewmen, and called to him.

"Hey, Bill, are any of the "Clan" here yet?"

"Yes, Sir, they're over at a table by the jukebox."

"Yeah, I see them now. Come on, let's see what they're up to. Hey, Nat, how long have you all been here?"

"We just got in a few minutes ago. Pull up a chair and listen to some of this good Glenn Miller music. If you don't like him how about Harry James or Tommy Dorsey?"

"I like all of them but probably Glenn Miller is my favorite, especially 'Moonlight Serenade' and 'String of Pearls.'"

"Captain, you said you had a surprise for us, and now we can't wait ... What is it?"

"Hasn't anyone told you about a special photograph in here?"

"No, I don't think so."

"Well, look on the back wall at the largest picture."

"Hey, it's some B-26 Marauders."

"Yeah, but go over and take a closer look."

"Doggone, that's our ship ... Nat, come over here, it's our plane flying close formation with Barnett's ship. When did they take this shot?"

"Do you remember last month when the squadron photographer was riding with Stangle? They tipped us off that he was looking for some close formation shots. When he signaled from Stangle's ship, Barnett and I moved in close and that's what he got, a whole series of close formation shots of our aircraft. One of the best photographs was enlarged and there it is."

"Man, that's a beautiful shot. Do you think we can get some copies?"

"Sure, I'll talk to Captain Jim Snow, our Group Photo Officer. He can probably come up with some extra copies. Anyway that's the big surprise.

"Boy, that's really something ... A shot like that could make it in Life or Time magazine."

"Well, they have sent some copies to the Air Force Public Relations Department so there is no telling where it may be printed later."

"You know, that might turn out to be a famous picture in magazines and books about aircraft."

"Yes, I guess it could, but there are a lot of photos being taken by Air Force photographers around the world."

"Captain, why don't you go with us tonight ... I think some of the bunch are going to catch a show at the Saville Theater ... It might be good."

"I hate to say no, but I'm worn out after all my sight seeing today. Hey, I nearly forgot ... I checked on train schedules today and there is a 4 o'clock to Chelmsford if you guys want to make it tomorrow. Sunday will be our last day in London, so what are all of you going to do tomorrow?"

"We'll probably do some more sight seeing or whatever the bunch agrees on."

"Okay, try and see the War Memorial in front of St. George Hospital or maybe visit St. Paul's Cathedral on Ludgate Hill. They say the choirs are beautiful there, and Sunday should be a good day to visit. You guys have a good time tonight and I'll meet you at the train station tomorrow."

"All right, Capt'n, we'll see you tomorrow."

The men stayed to talk for a while and Quinn headed for the hotel. He was more tired than he realized from walking all day, and after leaving a request for an 8 o'clock wake-up call for Sunday morning at the hotel desk he went upstairs to his room and quickly fell asleep.

The following morning the air was cool and brisk as Quinn made his way to Ludgate and St. Paul's Cathedral. As he approached several blocks away he heard the bells begin to ring a familiar melody. How beautiful to hear the full throated tone of the bells as they reverberated from building to building. Then the structure itself came into sight and he could see the magnificence of the building, so tall and stately. What a beautiful work of art. After climbing the front steps, he entered the huge marble floored entry and was inspired by its marbled pavilions and carved structures. He knew that Sir Christopher Wrenn had designed the cathedral and he stood awed by the thought that man had engineered such an exquisite structure. He entered the main Nave and looked up in amazement at the beauty of carved ceilings, colonnades and stained glass windows. A feeling of nearness to his God overwhelmed him. God's Spirit seemed to be touching him with the awe inspiring beauty of this house of God.

Farther down the aisle a large group of worshipers had gathered in hushed silence waiting for the service to start. Off to the left a Boys Choir began to sing a hymn, and the sound was exquisitely angelic. More than a hundred voices in perfect harmony was deeply inspiring. It was an experience of beauty and worship for Quinn throughout the entire service. After the last hymn and dismissal of Church he overheard someone say the East Wing was being repaired. Could the V-1 bombs have hit and damaged part of the Cathedral? What a sad thought to think of man's destructiveness, all set in motion by a power mad group of Axis Dictators. Surely this house of God would continue to stand, and German bombs would not be able to destroy it. He breathed a silent prayer, "Oh, Lord, please let this thy house remain through all this conflict."

How good it felt to be in God's house. It just always made the day go so much better, and today was no exception as he continued to see some of his favorite places in London. All the world seemed to be at peace, and he went through the day with a feeling of contentment.

The time drew close for his rail trip back to Rivenhall and he walked by his hotel room to check out and make his way to the rail station. The station area seemed crowded with servicemen all returning to their bases after a few days in London. He got his ticket and began to look for Zola and Nat. They were seemingly nowhere to be found, but it was a large station and he could have missed them somewhere. It was nearly 4 o'clock as he walked out to the track platforms from the main gate. He didn't have to look a long time for he could see a group of servicemen trying to board early for better seats on the train. As he boarded he heard Zola's voice, "Captain West, here we are. We've saved a seat for you."

"It's a good thing you kept one for me. I didn't know it was going to be this crowded. Did you guys have a good time this weekend?"

"Yes, Sir, we had a great time. How about you?"

"Oh, it was really good. It couldn't have been better. I'm real tired, but it's a good tired feeling, just body tiredness. Mentally I'm rested and that's what I needed was a break from the routine. Now I'm ready to go back."

"I think we feel the same way Capt'n ... It was good to get away from the field for a while."

It wasn't long before the train began to move slowly out of the station confines and into the open track of the suburbs. Then later, the men had all dozed off and were sleeping through the towns and villages until they were near the outskirts of Chelmsford and the trainmaster was calling Chelmsford Station, "All off for Chelmsford."

They began to slowly get their overnight bags together and wait for the lurching stop that would announce their arrival at the station. Then it was a few miles walk to the field unless they were lucky enough to catch a G.I. truck heading that way. As luck would have it, they caught two trucks going back to the air base and some tired, but also happy men would have no trouble sleeping that night.

CHAPTER THIRTEEN

COMBAT MISSIONS, CLOYES TO CAUMONT

The next morning the topic of conversation at the mess hall during breakfast was the "American Eagle in Britain" broadcast. Everyone wanted to know how the interview had turned out. Some had heard that it was a real good broadcast, but they wanted to know all of the details. Zola and Nat were answering most of the questions about it.

"Well, there were thirty or forty people there, mostly other crews that were being interviewed that morning. It was fun to hear that guy talk to the other crews. He asked them where they were from in the States and got them to tell their story of what had happened in their part of the ship. It was interesting. Nat, tell them about that guy in the B-17."

"Yeah, they had a bunch who flew 17's over Germany, and one of the gunners had shot down two German fighters in one raid. We had a good time. What about you guys here? How have the missions been going?"

"The weather has been terrible here while you all were gone. Missions have been put on hold, then scrubbed, then rescheduled. It has really been a mess. It looks like our mission today might be nearly the same way. Your crew are scheduled, aren't they?"

"Yeah, we're on the list, so I guess we'll be seeing you guys at the briefing this evening. See you later."

The briefing room was alive with chatter, as the men talked of the morning mission to Montreuil. There had been six aircraft hit and damaged by flak, but the bombing results were good.

Captain Morrow was giving his part of the briefing.

"The target is a railroad bridge at Cloyes. Weather over the target is 5/10ths cumulus. You have all your coordinates. Fighter escort at 1720. Take off at 1700. Good luck, men."

It turned out to be a near perfect mission with all bombs on target, excellent results, and only one plane damaged by flak. The next day they were scheduled again to bomb another railroad bridge at Epernon. There would be three boxes of twelve planes each and West was leading the third box. It was nearly a carbon copy of the mission the day before with good results and only two aircraft returning with damage. As the planes pulled into the hardstand area there was the familiar sight of Robbie holding "Jiggs" in his arms to keep him from bolting out to the aircraft.

Quinn wheeled in to his parking area and cut the engines. The men piled out and Jiggs was greeting each one with a wagging tail and a brief stop for his pat on the back, until he saw Quinn and started barking playfully.

"Okay, boy, you want me to play rough don't you? I know, come on, jump. Jump high ... That's right, good boy ... Jiggs, you're a dude aren't you?"

"Capt'n, that dog really loves you. He wants to follow you everywhere. I've never seen anything like it."

"Yeah, he's really something. If I had him at home I'd teach him how to hunt. He is plenty smart."

"When he is around here and your plane is gone he'll mind us pretty good. We can tell him to sit or lay down and he will look at us like he knows what we are saying, then he'll lay down. He's smart all right."

"Robbie, we may get a few days rest. If we get a new plane maybe that squadron painter can paint "By-Golly II" on her."

The few days off passed swiftly and on July 30 they were scheduled again to bomb a defended area near Caumont, France. The morning flight had bombed the same area with no flak damage. Now they were scheduled to bomb the same place in a late evening flight. The weather had been bad all week and if the mission wasn't scrubbed it would be a miracle. They later found out that it was a pathfinder mission so the group took off in 10/10 cloud cover and went over the target in 10/10 cover. The mission was classed as good so the "Gee" operators had done a good job of finding the target on radar, but the German gunner's radar had been accurate also as eighteen aircraft came back with flak damage.

Quinn's crew got a break the next day and weren't scheduled. They had gotten a baseball game together with some of the other crews. The ones that weren't playing were cheering for their teams. They had brought Jiggs up to the company area, which was a rare treat for him, and he had worn himself out scampering to catch grounders, running along beside the men who were running the bases. It was a good day and the men had enjoyed it all.

At the evening meal the men were tired but still laughing and talking about the game they played. Some were going to the movie and others were going by the Aero Club later, but for the most part, they would turn in early and get ready for tomorrow's mission.

Quinn had left the group a little early and headed to the Operations Building. He was duty officer that night and would be in on all the planning for the next day's mission. He also knew that he, Taylor, and McLeod were going to be the box leaders, so they would all be involved in the early planning.

Bob Curran giving Jimmy a box of oranges from his friends in the 397th Bomb Group, a rare item in Britain during WWII. Jimmy has a 598th insignia, a G.I. cartridge belt and as a final surprise was given a ride as co-pilot in a B-26 Marauder.

CHAPTER FOURTEEN

THE LAST MISSION

That evening the weather had begun to look a little better and the probability of a mission tomorrow looked good. Captain West walked over to the Operations Building and checked the loading list. "Yes, he thought to himself, there it is #4126-H. We're in the slot to lead the second box, and also they've got us listed as the camera plane. With a nine man crew and a full bomb load it might take the whole runway to get her airborne, but I won't need to worry about that. Those Marauders always make it, loaded or not."

He walked over to the mess hall hoping that he might run into Major McLeod or Captain Taylor and talk over some of the details of the mission. He wondered where the target area would be and he thought aloud, "Probably a railroad bridge around the St. Lo area." They had been hitting that section pretty heavy lately. St. Lo, it always reminded him of the pilots and crews that had been shot down in that area. It was something that you had to accept as part of war, but still it was a sickening feeling to think that lives were being lost. He thought of their parents and wives. How could they handle such a loss? It was frustrating to even think about it.

It wasn't like him to center his attention on circumstances over which he had no control, but the feeling of compassion for these men and their families had overwhelmed him. He turned his thoughts to other things and remembered that tomorrow was the 1st of August. Only ten more days until his 25th birthday. Several days ago he had gotten a package from home, and just as he suspected it was a present from his mother. She always remembered his birthday and wherever he was, a package always made its way to him.

The mess hall wasn't very crowded, and when he saw McLeod at one of the tables he went over and took a seat beside him.

Quinn asked, "What time are you planning our briefing session tonight?"

"Well, whatever time suits you guys best. What about 8 p.m.?"

"Sounds okay to me. Did you see the orders from 98th Wing?"

"Yes, they came in a few hours ago."

"What's the target?"

"A railroad bridge over the Loir River a few miles west of Angers at Les Ponts De Ce. It shouldn't be too bad. There hasn't been much flak activity in that area lately."

"Okay, I'll see you at 8."

Quinn walked to his quarters musing to himself, "Not much flak, that would be a relief." It seemed like most of the missions lately had been to hot spots where the aircraft had caught plenty of flak. One good sign was that German fighter activity had seemed to be fading out with each mission they flew. The flak was bad enough, but they could dodge around some of it.

This would be the 39th mission for the By-Golly crew, and that meant only 11 more missions to pull before they would be eligible for a leave home and stateside service for the rest of the war. He thought about the missions and counted them just like the men did, but he never would let himself get as outspoken as they did about it. He knew it was necessary for him to show more confidence and reserve than others in the crew. The men needed someone to look to for confidence, and he tried to meet that need, but he was sweating the missions just like they were. He laughed as he thought about Zola's exuberance after each mission as he would slap Nat on the back and yell, "Number 38, Nat. Sweat 'em boy, sweat 'em." They had it on the down hill pull now. Only eleven more to go. He shook his head and thought to himself, "Get out of this rut man, there's work to do on the Angers mission. No time to be daydreaming about going home."

He looked at his watch and headed over to the pre-briefing with Major McLeod at Group Headquarters. Captain Taylor would probably be there by now, and the duty officers who would help plan the mission would be along soon. Quinn walked into the small conference room and sat down as Major McLeod spoke, "Pour yourself some coffee, West. It looks like this will be a long one."

The tables were covered with navigational maps of the Angers area, weather reports, orders from Wing and Group Headquarters, and reports of weather conditions over the target area. McLeod was talking to the navigation duty officer, "Time over target is 0900, so work back and give me an estimated take-off time. We've got a south-southwest wind at 10 knots, and an average airspeed of 210 MPH. Captain West, you will be leading the second box of 12 aircraft, so you'll want to get your navigator and bombardier to keep a close check on headings and wind drift. The whole box will drop their bombs on your signal, okay?"

"Yes, Sir, what about bomb load?"

"You'll carry four 1,000 pound general purpose bombs, fused at 1/100 delay in nose and 1/40 delay in tail. That reminds me, Lt. Coyne, call the armament crews and give them the bomb load and fuse settings."

Major McLeod looked over some of the photographic work done preceding the mission and then spoke, "Okay, men, here it is, the Les Ponts

De Ce Railroad Bridge. Three spans, concrete reinforced with steel girders. She will be tough to knock down unless we get some direct hits on each span. The bombing patterns from each box will have to be near perfect. Time over target is set for 0900, fighter rendezvous at St. Catherine's Point. Lt. Bown, what times are you figuring for take-off?"

"Wake up call at 0500, briefing at 0600, engine start at 0645, take-off at 0700, and rendezvous with fighter escort at 0720."

"All right, hold onto the navigational maps and headings. I want you to present them at the 0600 briefing. Our weather doesn't look good over target right now. It's solid overcast, but they expect some clearing to about 2/10ths cumulus at take-off time. Are there any questions? Okay, get a good night's sleep and get ready for an early mission tomorrow."

The group left the headquarters area and headed to the Officer's Quarters. It was late and Quinn was tired. There had been so many late night planning sessions and early morning missions that it had begun to wear him down. He was glad to hit the sack and try to sleep a little before wake up.

Sleep didn't come easy that night for some reason. He was restless and woke earlier than the others. He dressed and walked to the mess hall hoping to find out about whether the mission had been scrubbed because of weather or if they would fly it as scheduled. He spotted the CQ in charge of wake up and found that the mission had been scrubbed. It was a relief for him. At least the other men would not have to be awakened this early. He got an early breakfast and had almost finished when Major McLeod came in. "Good morning, West, did they tell you the mission had been scrubbed?"

"Yes, Sir. What was the reason? Weather?"

"Yes, it is 10/10 cloud cover over the target, so we have rescheduled for a mission at 1500 take-off time. We will need to have a short prebriefing after noon chow today. That is if the weather clears over the target in time."

"Well, I guess that will give me some time to finish a few reports and get some of the operations work done. I'll see you this afternoon."

Quinn had been promoted to Operations Officer of the 598th Squadron a month ago, and the work was heavy. There were clerks at headquarters to type the directives, but the planning and working out problems was time consuming. Major Bronson had been the Operations Officer until his plane and crew got shot down over France. Bronson was from Memphis, Tennessee, and he and Quinn had talked many times about places in town that they knew about. They had often frequented Loew's Palace and Strand Theaters, and after the show, some food at the Peabody Grill or Jim's Restaurant. He always liked Bronson and it hurt deeply to lose his friend.

Later that afternoon at the pre-briefing it was observed that the weather was clearing. The times were set for briefing, take-off, fighter rendezvous, and time over target. The first box would be over target at 1706, the second at 1707, and the third box at 1708. Everything was precision and accuracy. There was little room for mistakes.

The final briefing was set for 1430, and as the men began to arrive their eyes were drawn to a large map of France in the briefing room. The red string on the map extended from Rivenhall to Angers and retraced a different route of return. Areas of expected heavy flak were circled, and with a quick glance the men could tell what they would be up against.

Col. Coiner was the first to speak, "Men, this is an important railroad between Angers and Paris, and there is plenty of supply materials going over the Loir by this route. We've got to take this bridge out fast. It might be a key to bottling the German's retreat from Normandy and possibly shortening the war. Major McLeod has the statistics for the main thrust of the mission."

McLeod began his talk, "Take-off at 1512, rendezvous with escort at 1542 St. Catherine's Point, time over target 1706. Here are all your coordinates. Time over IP is 1715. After drop, turn left and retrace route back to base. Weather is good over target 2/10ths cumulus, visibility 3 to 4 miles upsun, visibility downsun unlimited. Bomb at 12,000 feet altitude. Cloud conditions at take-off are solid overcast changing to excellent over the Channel. Notice the line here past Normandy. This is the advance line of our troops. If you get hit, make every effort to get back past this line away from enemy territory. Good luck, men. Engine start up time is 1500."

The men began to leave the briefing room and head for the jeeps and trucks that would take them out to their planes on the hardstand areas. Already the armorers had loaded the bombs into the bomb bays of the planes and set the fuses. The machine gun ammunition had been loaded into racks above the gun positions.

Engines had been checked by the crew chiefs. Control and radio checks had been made. All was in readiness for the flight crews to arrive and make a last check of equipment before take-off.

Quinn and Budge had gotten to their plane before the crew arrived. West saw Robbie checking a tire, and he called out, "Robbie, did she check out okay?"

"Yes, Sir, she's fine and ready to go."

"All right. Here comes the crew. Okay, Zola, get up in there and look her over. Nat, Pic, Brinn, climb in there men. Okay Cramer, where's Daoust? Here he comes. Budge, go on through and I'll follow you. Here comes Webb. Lloyd, hurry up, a few more minutes and we would have left you," Quinn laughed. Budge and West moved through to the cockpit area and Captain West took the left seat and slid back the small window to talk to Robbie. "Robbie, we're ready to hook up the external power." Captain West began to check off the engine starting procedures. Cowl flaps open, check fuel gauges, set throttles, master switch on, left booster pump on and prime left engine, energizer switch on left. The inertia flywheel began to whine as it picked up speed and then the mesh switch was moved to the left position and the propeller began to turn slowly at first, then firing intermittently with clumps of white exhaust pouring out the stacks as she caught and began to run at 800 RPM. Then the right engine started and Quinn signaled to Robbie to pull the external power and clear away from the plane before she started her taxi run.

Robbie gave a signal of "Okay," and Quinn pulled the plane out slowly to the taxi way.

The engines sounded good and Quinn eased them up to 1000 RPM, heading out the taxi strip to the main runway. When nearly there, he angled his plane to the side of the runway and ran his engines to 51 inches of manifold pressure and 2700 RPM; checking fuel pressure, oil pressure and temp, cylinder head temp. Everything looked good. He checked his watch and plugged into the intercom. "Men, we're one minute until take off." The men all knew what he meant. It was an unspoken request for prayer for a successful mission.

Budge had already called in to the control tower, and West was looking over that direction waiting for a green flare that would be a signal for take off. "There's the green one, Bill, let's go." The aircraft began to accelerate quickly, 120, 130 MPH, and finally the nose wheel had come off the ground but the mains were still heavy and seconds became minutes as the runway had nearly been eaten away completely. A final pull on the control column and she lifted grudgingly and slowly as the boundary markers whizzed past and the plane barely cleared the tree line at the far end of the airfield. "Pop the gears, Bill, we're home free now. That was a close one. A nine man crew makes a difference, doesn't it?"

"Yes, Sir, we're loaded heavy today."

They gained altitude and began to circle the field to form up with the other aircraft, then they headed out over the countryside of Essex and on to the Channel after picking up their escort at St. Catherine's Point. Zola had seen them first as he talked over the intercom, "There they are Captain, P-51 Mustangs at 3 o'clock high."

The weather reports had been right, for they had encountered a solid overcast up to 2500 feet then it became scattered. After reaching the Channel, it was unlimited visibility. They continued to climb over the

Channel to 8,000 feet as they reached the French coast. The men had cleared their guns and now it was a waiting game until reaching the coast and beginning the evasive turns to avoid enemy flak. The gunners searched the skies for enemy fighters and Cramer and Daoust checked their calculations looking for the coast and landmarks that would tell them that they were on course.

The intercom came on as Daoust called out, "Captain, begin to take evasive action." The pattern continued as West followed Daoust's directions and still maintained a constant altitude, speed, and attitude of the plane ready to roll out on the prescribed course just before reaching the IP, where a straight and level course must be maintained until the bombardier signals "Bombs away." Daoust came over the intercom again, "Captain, left 2 degrees, hold, we're over the IP ... Target sighted, bomb bay doors open, hold her steady." Zola's voice broke in over the microphone, "Captain, enemy fighters coming in at 3 o'clock high. It looks like two FW-190's. Pic, heads up ... Here they come. Nat, are you on a waist gun?"

"Yep, I'm ready ... They're both coming straight in on us."

The rattle of machine gun fire sounded from Zola's guns, then bursts of firing from Nat's guns at the waist section, and finally the bursting of 20 mm cannon shells from the fighters which were ripping into the center section and left engine of the aircraft.

Daoust's voice came over the intercom, "Capt'n, hold her level. We're only 20 seconds to target."

"Pic, there's two Messerschmitts coming in at 6 high ... They're lining up for a run from tail to nose."

"I see 'em, boy ... They look like they are trying to ram us."

Pic's guns began to fire again and Zola's pair of fifties began to add to the den of firing and smell of burnt cordite.

"I got him," Zola shouted, "I got him ... He's smoking."

Nat's guns in the tail section began firing at the second Messerschmitt, and Zola's guns answered in quick reply. Both enemy aircraft had pulled up high before getting any good hits on the Marauder.

"Nat, they're pulling high and breaking left ... I think we missed them. Dog gone, I was sure I had hit the first one."

"Did they break left into the cloud bank?"

"Yeah, maybe they'll pick on another plane for the next run. Where are those Mustangs that are supposed to be covering for us?"

"I don't know ... Some of them peeled off and headed down to ground level to strafe ... There's still a few around here somewhere."

"Pic, they're coming in again at 6 level ... They must want us pretty bad

... Heads up, boy."

"Nat, there's one of the Mustangs after the Messerschmitts ... He's got him! Boy, he's got him! The Messerschmitt is falling and on fire."

"Good, that's one less we have to tangle with."

"Where is the other bandit?"

"He broke left real fast when that P-51 got after him."

"Pic, they're not through with us ... Here comes a pair of BF-109's at 6 o'clock high to level, boring in fast. Looks like they're heading straight for the tail section."

"I've got 'em in my sights now."

Pickelsimer's guns began to fire several long bursts, then Zola's guns picked up the firing as a hail of bullet hits from the BF-109's were peppering the top center section of the Marauder. Another Messerschmitt passed over high firing tracers at the plane all the way across, but not getting good hits. The gunners in the Marauder were answering fire for fire and the rattle and explosions of all guns firing was deafening.

"He missed us Zola ... Did you hit him going over?"

"I got some hits, but not enough ... Nat, keep after them ... They're coming in again at 10 o'clock level ... It's the FW-190's ... That cannon fire is deadly, boy!"

The Focke-Wulf 190's were coming in together, bent on the destruction of the Marauder. The enemy knew the importance of the lead ship. If they could shoot down the lead aircraft then the bomb drop would be scattered and ineffective. They pressed in with a frontal attack trying to rack the cockpit area and both engines with cannon shells.

The Marauder was like a sitting duck not being able to move off course or take any evasive turns. It was 10 seconds till target and West had to maintain a steady course for the bomb drop to be effective. The whole mission depended on his steadiness under fire. To turn right or left now and evade the fighters would be a violation of everything he had been taught as a Marauder pilot. He was a lead pilot and he was expected to react with steadfastness on the bomb run.

The enemy fighters cannon shells began to spatter and explode everywhere. Nat's and Zola's guns were firing incessantly, but to little avail, as the fighters were zeroing in on the aircraft, their hits were starting fires in the cockpit area.

The Marauder lurched sideways and lost altitude as shell bursts were hitting the left engine. West yelled over the intercom, "Daoust, pull the salvo lever and get rid of the bomb load ... We're hit bad and on fire. Bill, move back so Daoust can come through."

"I've been hit in the knee by flak, Capt'n ... It's bleeding pretty bad."

"Here, I'll help you. Daoust are you okay? Come on out of there quick ... This smoke is terrible. I can hardly see. Bill, get ready to bail out ... I'm hitting the bail out alarm."

It was the dreaded sound for all Marauder men, as the bail out alarm sounded like nothing else on the face of this earth. They had heard it in practice sessions and it was enough to chill their blood. Now it was even worse coming through the intercoms and reverberating through the entire ship ... It's incessant sound was the last death rattle of a dying Marauder. It was abandon ship without hesitation. To hesitate was fatal, as they only had a few seconds before the plane might go into a dive, spin, or snap roll that could pin them all to the wall of the ship and they would never get out.

West was fighting the controls trying to keep the ship from yawing left. The left engine was burning now and the right engine was trying to pull the plane into a left spin. He had let the gears down so the men could bail out through the front wheel well and this was making the ship harder to control. Budge had jumped out first, so Daoust would have room to come through the crawl space from the bombardier's compartment. Daoust had come through seconds later burned badly on his face and hands. The whole ship was now filled with dense smoke. The men who were left in the ship were choking and becoming dizzy from the noxious fumes. West was fighting to remain conscious, and he had managed to put the plane on auto pilot with enough trim to hold her steady so he could help some of the men get out.

"Come on Nat, where's Zola? Come on, Pic, where's Cramer?"

Pic answered, "I thought I saw him getting his chute pack back there, but it could have been Zola, I can't see a thing."

"Zola, come on ... Get out. Hurry ... Get out!"

All had made it out in the nick of time; coughing, dizzy, and nearly blinded by the fumes. There were seven parachutes that opened as they cleared the burning aircraft.

Quinn's breath was coming in short gasps as he tried to keep from breathing the heavy smoke. Everything was dizziness and confusion. "Cramer?" West was fighting to stay conscious. He thought, "If I can get to him I'll push him out and pull his rip cord, then I'll jump myself. Cramer, where are you?"

Nothing seemed clear anymore ... The walls of the plane were fading in and out, and the smoke was overpowering. West wanted to save Cramer, but he was weakening.

His last thoughts were of trying to see that all would get out safely. "Cramer? ... If I could only ... If I could ..."

The plane was on auto pilot and circling left, losing altitude. The left engine was in flames now, and the last two members of the By-Golly crew were lying on the floor of the plane, unconscious from smoke inhalation.

The crew who had bailed out couldn't see the plane now, as it circled out of their range. The other Marauders of the Group were heading back to their base in England, and only two Mustangs were left fighting it out with the German fighter aircraft. The Mustangs had shot down one of the Messerschmitts and the other German pilots broke off the combat and headed northeast toward their base.

As the P-51 pilots left, they saw the crippled Marauder circling close to the ground, and then burst into flame as it hit in an open meadow.

On the ground the explosion was deafening and it had reverberated for miles through the French countryside.

The cattle that were grazing in the field started to run in confusion, then stopped, not knowing where the noise had come from ... Then a quietness settled in the vacuum following the great noise, and everything seemed to pause in a moment of stillness.

The cattle which had bolted when the explosion occurred had settled down now and were grazing contentedly. On the edge of the meadow in the woods the birds had begun their plaintive songs, and all was at peace.

For Captain John Quinn West, the war was over. No more struggle and preparation for war ... Only peace ... No more tumult and shouting of battle ... Only the quietness of peace ... God given peace.

In the distance the faint rumble and thunder of aircraft of the 397th could be heard returning to their base. For these men the war was still a reality ... Something to contend with ... Something to fear ... But, for West and Cramer, the war was over.

Quinn had wanted so much to be home for Christmas, and now he was in his real home ... The eternal mansions of Glory.

For those loved ones he had left behind, only the comfort of their Lord's words could still their deepest sorrows.

"Let not your heart be troubled; ye believe in God, believe also in me.
In my Father's house are many mansions: if it were not so,
I would have told you. I go to prepare a place for you.
And if I go and prepare a place for you,
I will come again and receive you unto myself;
that where I am, there ye may be also."
   John 14:2

CHAPTER FIFTEEN

THE BAIL OUT AND ESCAPE

When the bail out alarm sounded, Budge had struggled to get his chute on. Everything was confused as the flight deck filled with dense smoke. He had to clear out in a hurry so Daoust could come through the crawl space. When he tried to stand his knee buckled. He was hurt badly and bleeding from a shrapnel hit in his leg. He steadied himself by holding onto the bulkheads and walking back along the passageway to the open bomb bay doors, and then jumped.

The pain was nearly unbearable as his chute opened with a hard jerk. Now he could see two chutes below him, so some of the crew had gotten out before him. He strained to get a view above, but his canopy blocked out most of it. As seconds passed he could see the aircraft was still burning and moving far away from the area where they would drop. He dreaded hitting the ground and knew there was no way to keep the fall from splintering his bad leg. The chute was slowly drifting towards a clearing near some woods. Then it happened, as the ground came up faster and faster, he hit and it was difficult to keep from blacking out. He fought against it knowing that he would have to crawl to the woods if he wanted to escape the German troops which would soon be searching for them.

He clenched his teeth and began to crawl, dragging his bad leg, but knowing he could never make it to the woods.

Then he heard a voice, "Budge, we're over here."

"I can't move, Nat, I'm hurt bad. I think my knee is broken."

Nat and Pic started dragging him toward the woods until they heard a motorized vehicle approaching.

"Wait a minute ... Dog gone, it's a German command car."

"How many guys are in it?"

"Six, and here comes another car ... We're sunk."

"Where are Daoust and Zola?"

"Daoust is burned bad and Zola is over there trying to help him, but none of us had time to hide anywhere."

The car stopped a stones throw from them and the troopers jumped out carrying automatic weapons. One of the men spoke in near flawless English with a slight German accent.

"Come out and keep your hands up."

Nat and Pic began to slowly move out from a low spot that had partially hidden them.

"One of our men is wounded."

"Bring him out with you."

Nat and Pic got Budge between them and helped him walk out.

"Where are the others?"

"We're not sure."

The Germans conversed and one of the cars began to search the area and found Daoust and Zola. They loaded them into the cars as well as they could, placed a tourniquet above Budge's knee to slow the bleeding, and headed back to the town. The ride back to the French town was long and over rough terrain, and as they drove through the town and stopped in front of a large building, Nat asked the German officer a question.

"You aren't going to separate us ... are you?"

"The ones who are hurt go to the hospital ... The others go to the prison compound."

The German officer disappeared into the building and returned with two men and a stretcher. The other car had pulled in and Daoust was led into the hospital entrance.

The two cars started and drove to the edge of town and entered a large military post with barracks and drill fields, as they drove further there were barracks fenced with barbed wire that definitely looked like a prison compound. The cars stopped at the gate and unloaded. A German guard opened the gate and let them through, then closed and locked it. They were all searched and told to bunk in Barracks Six. As they walked toward the barracks a whistle blew and they were told that it was a signal for fall out for inspection.

While they were in formation, instructions were given for work details and they were told about punishments for attempted escape. Nat happened to see a familiar face in front of him.

He whispered, "Skarles, where are you guys sleeping?"

"Over in seven."

"How many of the outfit are here?"

"Not many ... Some of them got shipped out last week."

Later, after formation, Nat and Skarles got to talk."Where are they shipping these guys?"

"The rumor is to Germany."

"Boy, I sure don't want to go to Germany."

"They are afraid this country will fall into American hands soon and the prisoners will be freed."

"Skarles, has anyone ever escaped from this compound?"

"I don't know ... I haven't been here that long."

"Well, it don't look like a very tough place to me."

"Yeah, but you could get shot trying to escape or get in solitary if they catch you."

"I know, but what are they doing to those guys in Germany? Maybe they're shooting them anyway. I don't like the idea of getting shipped out. Try to think of something and we'll talk again."

The days moved slowly with work details during the day and cramped quarters at night. Zola had been moved to Number Three, and was taking care of Budge and Daoust who had been released from the hospital and moved to the compound.

Several weeks later there was a rumor going around the camp that Paris had been liberated by American and French troops. It was probably true because the Germans began to get everything packed for movement to another camp. Everyone was afraid they were all going to be moved to Germany.

Nat and Skarles got together again and secretly planned an escape. They met at Skarles' barracks and talked in whispers.

"Skarles, have you noticed how lax now they are getting about counting prisoners?"

"Yeah, if we could slip off from a detail, no one would notice."

"I'm thinking the same thing... Listen, tomorrow if we go on our regular detail, when we load into our truck going back, look for that first wooden bridge. Just after it, there is a grassy bank we can jump into and they'll never miss us."

"All right ... We'll do it."

The plan worked and Nat and Skarles hid under the bridge until dark, then they began to walk toward the American lines and freedom.

Budge and Daoust had stayed at the camp because they had not fully recovered from their wounds during the crash. Zola and Pic didn't want to leave them in their condition, but the day came when they would have to be left behind. The camp was ordered to move and all prisoners would be forced to leave except those who could not walk, so Budge stayed and was freed soon thereafter by American troops.

Zola, Pic, and Daoust made the trip to Germany, but it was not long after that until the American troops had moved into Germany and they were freed and allowed to go back to their units. Nat and Skarles had gotten back early, and by then the 397th had moved to bases in France.

After getting back to their unit, the first information they wanted to know was, had Capt'n West returned, where was Robbie, and was Jiggs still there? Jiggs was still there patiently waiting for the return of his friends. Several aircraft had been serviced by Robbie's crew, and Jiggs liked for all the crewmen to pet him, but the real spark had somehow gone. Robbie didn't have to hold him now when the planes came in. Jiggs would watch each man as he came out of the plane, but none seemed to be the one he was looking for. He would just lay down and wag his tail slowly as each one passed and petted him. However, the return of Nat brought an excited greeting from Jiggs, and the reunion of Jiggs, Nat and the ground crew lifted the spirits of everyone.

Nat and Robbie talked for several days trying to catch up on all the news.

"Robbie, have they heard anything about Capt'n West?"

"Not a thing, and it's killing me. I just can't figure why he hasn't shown up yet."

"If he and Cramer were captured it may still be a long time until we find out because they are shipping prisoners deep into Germany and it won't be until after the war before we will really know what happened to them. Don't give up hope, Robbie."

"Don't worry, I never quit thinking about you guys. Hey, that reminds me. Someone saw Budge in London and said he was being shipped back to an army hospital in the States."

"Yeah, I thought he would be one of the first of our crew to get back because of his injury. Those German doctors did some good work on him, but he was still bothered by pain when he walked on it too much. I guess it takes a lot of time for something like that to heal."

"Have you heard anything about when your orders will come for you to go home?"

"Skarles said he had found out that it would be just a few more days. I'm sure going to miss you guys. I wish I could see Zola and Daoust and Picklesimer again, but there is no telling when they might be released from P.O.W. camp. It could be months before they show up."

It was only a matter of days before Nat and Skarles shipped out, and Robbie and his ground crew got down to the business of taking care of the Marauders returning from combat missions over Germany. There was always plenty to do as the missions were still difficult with flak damage to repair as well as loading bombs for each new mission. It was a never ending struggle, and the days drug on through September and October.

One morning most of the men had finished breakfast when they heard some commotion in front of the mess hall. Several guys were yelling and laughing as three men walked in the front door. It was Zola, Pic, and Daoust coming in with some of the guys in their old squadron. They had been liberated from one of the German P.O.W. camps and sent back to their home outfit. Soon everyone was all around them asking questions about where they had been and what had happened to them these past few months. It was a happy gathering with old friends and buddies all trying to jump into the conversation at once.

Zola told them that they had tried to stay with Budge at the prison camp in France, but they had been forced to move to a camp in Germany, then later the Americans had liberated their area and they were allowed to go back to their old units.

"Zola, we heard that Budge had been sent back to the States because of his bad leg and we didn't get to see him. Hey, I nearly forgot, you knew Nat and Skarles escaped didn't you?"

"Yes, we heard in camp that they had tried to escape, but we never knew whether they had been captured again or what."

"They made it okay, and about two months ago they came walking into camp and they looked just like always, laughing and joking about their escape. They got shipped out during September, and I guess they are home by now."

"Boy, I wish I could see those guys again. I'm really happy that they made it back home."

"I'll tell you someone who will be happy to see you guys get back and that's Robbie."

"Oh my gosh, I thought Robbie would be shipped out by now."

"No, he and the crew are still down there working on the Marauders in our squadron. Get one of the men to take you down in the truck. It's not far."

The squadron hardstands and armorer's area were only a few minutes from the mess hall and they saw Robbie as they pulled in.

Robbie yelled when he saw them, "I'll be dog gone if you guys aren't a sight for sore eyes. Boy, I can't believe it. I have thought about all of you every day since you've been gone."

"We've missed you too Robbie, it's really good to see you."

"Hey, here's another crew member who wants to welcome you back."

They all saw Jiggs run from behind the armorer's shed out to greet them.

"Man alive, look at Jiggs wag his tail. He hasn't been this happy in months."

"Do you think he remembers us?" Zola asked, and reached down to pet him.

"Sure he does, can't you tell how excited he is? He's jumping all over you. If you guys could have seen how down he was when you all didn't come back, you would know that he knows you are back and he's loving every minute of it."

It was like a tonic for the men to pet and "rough house" with Jiggs. They realized that they had also missed the companionship they had with Jiggs.

That night at the mess hall, McGinnis asked them about West.

"Zola, do you think West made it to a P.O.W. camp?"

"I don't know Neil, but I'm not leaving this place until I find out. When the Germans surrender, then all the prisoners will be freed, and I want to be here to see Captain West come home. I'm not leaving until I see him."

"I don't blame you one bit. I'd feel the same way. Hey, do you remember Ryherd's crew? Well, they went down on August 4, and you won't believe this, but Tiny Thorp came walking in on August 30. The Germans never caught him and he walked half way across France, and he's still flying missions. They tried to send him home, but he didn't want to go."

"That guy is really something, isn't he? Tell me what the group has been doing since we left?"

"Well, it's a long story, you remember we were about to move to Hurn when you were here. Bournemouth was beautiful, but we only stayed three weeks. While we were there, the group flew its 100th mission. They shot flares everywhere when the planes came in and nearly burned up the field. Then we moved to Gorges, France, and stayed there for just two weeks until we moved to Druex. We were at Druex for four weeks and moved to Peronne, France, and we've been here ever since. Now the group has flown 146 missions and I have 63 and can't get out of the rut to get 65 and get home. What a life."

"Yeah, I know what you mean. See you tomorrow, Neil."

The missions continued on days that they could fly, but winter weather was starting to be a problem with rain and mud, then snow, fog, and overcast days followed one upon another and it was taking forever for a man to build his mission total to 65 so he could get sent back to the States.

Neil's old buddies were all getting shipped home. Mitchler left December 14 and Schubin left on the 18th. It was really getting lonesome around the barracks with all the old crews gone and new men coming in every day to be replacements. He was sweating out his last two missions and hoping every day he would be scheduled quickly and get his last missions over with.

On December 23 he was scheduled for a mission to Malmedy, and Eller, Germany, to destroy a railroad bridge there. They were unable to make contact with their fighter escort but they proceeded on to the target alone where they encountered heavy flak just before reaching the target. One aircraft got hit in the right engine, rolled over on her back and crashed. A few minutes later another plane got blown in half. They made the bomb run and seconds later they got jumped by twenty-five German fighter aircraft. They were coming in on the first box as McGinnis was watching from his position in the second box. The enemy fighters scattered the B-26's, as the bombers took evasive action, but the fighters were riding them down with guns blazing. One fighter chased a Marauder until the B-26's left engine caught fire and it began falling. The tail gunner on the Marauder stuck to his position and shot down the fighter, but the Marauder gunner went down with his ship. Another Marauder top turret gunner did the same thing. He shot the fighter down, but went down with the crashing Marauder. The enemy fighters shot down seven of the bombers with a loss of three of their fighters. It was a bad exchange although the bridge was destroyed. The 397th received a Unit Citation for their bravery under fire.

On Christmas Day, McGinnis flew his 65th mission east of Luxembourg. Two ships were shot down on this run, and when the other crews returned and landed, a ship blew up on the runway. Then later they discovered that one of their planes had blown up on the take off run.

When McGinnis was gone, there weren't many of the old gang left except Zola and Robbie. Zola was trying very hard to track down the whereabouts of Captain West, but it was a difficult job to talk to officers and ask them to check the return P.O.W. lists from so many camps. He refused to think West had died in the crash. He thought it just couldn't be true. West was too good a man.

It was after the surrender of Germany in December of '45 when a headquarters orderly told Zola that Col. Coiner wanted to see him. Zola knew that now some word must have come, because he had gone to Coiner's office several times to beg him to call again and search the missing in action reports. Zola walked in and saluted. Coiner invited him to have a seat.

"Sergeant, I know you have been wanting to know what happened to West and Cramer. The German records are just now being given to us about the whereabouts of airmen who crashed or bailed out in enemy territory, and I don't know how to make this any easier for you, but West and Cramer died in the crash August 1st. The plane exploded on impact and both officers were killed."

Zola was trying to blink back the tears, as his mind was reeling and trying to recover from the inevitable truth of what Coiner had told him.

"I'm sorry, Sir."

"I am too, Zola, he was a fine pilot and a credit to his unit. I have been putting off writing to his wife and giving her the official notice, but I must do it soon. Do you think you might try to see any of them after you get back to the States?"

"I don't know, Sir. I don't think I could do it. Our whole crew thought so much of him."

"I understand."

"Colonel, there is one thing I would like to do if it's possible, and that is to get his dog back to his son in the States."

"That's a tall order, but you have my permission. I'll write a letter of intent for you, and hope you are able to cut through the red tape. Getting the dog out of France and England will be tough enough with all the quarantines, but getting him aboard a troop ship will be near impossible. I admire your wanting to do it though. I'll have your letter tomorrow. Good luck, Zola."

"Thank you, Sir."

Zola left and went to the mess hall looking for Robbie. He found him talking to some of the men in the squadron.

"Robbie, you and I will be leaving for home soon when they cut our orders. Do you think we might be able to get the Captain's dog home to his son?"

"Yeah, we'll sure try. What did Coiner say?"

"The records show that Captain West and Cramer died in the plane crash."

"I'm sorry, Zola, I think we felt it all along, but just wouldn't let ourselves think of that possibility."

"Yeah, you're right. I hate it though. It doesn't seem right leaving here without all the crew. Nat and Pic are gone, Budge and Daoust are gone. That's why we have to bring Jiggs back. He's the last one in our crew."

CHAPTER SIXTEEN

GOING HOME, 1945

The days were long waiting for orders to go to London and then to a port of debarkation for passage home, but the orders were finally posted, and Zola packed. He and Robbie and Jiggs got transportation to London. It was in London that they lost Jiggs somehow, and after frantically looking they found him in a dog pound there. It was necessary to get a lawyer to secure his release, and they were required to take him directly to a veterinarian for shots and quarantine. They had a few days before getting transportation to the port area, so they could make it.

They got Jiggs and all his papers and made it to the port and army separation camp. It was difficult, to say the least, but they got all the way to walking on board the troop transport going back to the States when immediately one of the ship's crewmen told Zola to report to the ship's captain. When he got to see the captain, the ship commander talked to him in a very gruff manner.

"You know there are regulations about bringing animals on board ship."

"Yes, Sir, but all his papers are in order and his quarantine has been completed."

"I'm sorry, no dogs on this ship, soldier."

"Captain, please let me explain. This is my pilot's dog. My pilot, Captain West, was killed in action after 39 missions, and I want his three year old son to at least have something that belonged to his father."

The captain thought for a minute and answered hesitatingly.

"Sergeant, if you promise to keep that dog off the deck, take care of him by yourself, and not bother my crew with him, you can keep him aboard."

"Yes, Sir, I really thank you, Sir, I promise you won't have any problems with him."

Zola left and went back to tell Robbie.

"Robbie, we'll have to keep him here beside our bunks. When I go to chow you keep him, so there will always be one of us to take care of him."

"Okay, but I can't believe that captain let you keep him on board. I thought it was all over for us."

"I know what you mean, I can't explain it either. That guy was tough as nails until I told him about Captain West. Then he seemed to change and told me to go ahead and keep the dog. I didn't believe it myself."

"Zola, do you think Jiggs might get sea sick?"

"Oh my gosh, I didn't think of that. Maybe he won't and I hope we won't either."

The trip was not as bad as they thought. They were able to get Jiggs some food scraps from the ship's galley each day. He even turned out to be a good traveler. They had found a wooden crate, and were able to make him a place to bed down and rest.

The days passed and their ship came to port in New York harbor. They were all excited and anxious to get their separation papers and discharge, but it was a lengthy process of physicals, getting separation pay, and checking in army equipment. They discovered that Jiggs had to go to a quarantine clinic for a week before being released, but they were able to handle even this.

After Jiggs' release, they arranged to have him shipped by rail to Sardis, and they told the shipping clerk to make him comfortable and ship him in the biggest box he could find.

They wrote to Captain West's family and told them to expect a freight shipment of Captain West's dog. The family waited for days, looking every day and calling the freight station at Sardis to see if the dog had arrived. Finally they got a call from the station. The agent told them, "Your dog is here."

They all went down to the station and what a large box it was. It was a box which was ordinarily used to ship pianos. The box was opened and Jiggs ran straight to Quinn's son, Johnny, wagging his tail enthusiastically. Jiggs seemed to have a sixth sense in realizing that Johnny was somehow a part of the master he had once loved months and months ago and then had lost.

In Jiggs' excitement he jumped up and licked Johnny in the face. Johnny laughed and wiped his face with his sleeve, then he bent down and petted Jiggs, ruffling the dog's coat and loving him much like Quinn used to do. Johnny could hardly wait to get back to the farm with Jiggs and run and play with him through the open fields.

When they arrived home Johnny called to Jiggs and they ran out through the gate to the back pasture. Jiggs was happily running and jumping beside him. The family watched with tear filled eyes until the pair had gone out of sight. They could tell that the bond of a boy and his dog would help to heal the hurt that some day the boy would feel knowing that his dad had been lost during the War.

And who can tell? Maybe Jiggs knew he had found a new master much like the one he had known in a far, far away place called England.

THE FAMOUS MARAUDER PHOTOGRAPH

The photo of the "By-Golly" flying close formation with Capt. Barnett's Marauder has been famous indeed, as a score of photos have been discovered in various places, magazine articles and in many books. The following are a few places and articles where the photo has been displayed.
1. Resides in Air Force records, Photographic Depository, Arlington.
2. Placed in London's USO Rainbow Club during WWII.
3. Printed in Newsweek, 1944.
4. Used in WWII exhibit at Smithsonian Institution.
5. Copy in book, FlakBait, Devon Francis.
6. Aero Album #3 by Ken Rust.
7. 397th BG History, Edited by Henry C. Beck, Jr.
8. Magazine, Air Classics, Diary of Marauder Crewman, McGinnis.
9. Saga of By-Golly, Stovall, Aviation Review.
10. When Marauders Flew for Hollywood, Stovall, Air Classics.
11. Air Combat, AAF Bombers at War, photo in article, Stovall.
12. Marauder Men, Stovall, Air Classics.
13. Air Force Magazine, 1944.
14. Rivenhall, The History of an Essex Airfield, B. A. Stait.
15. Wings of Courage, Stovall. Photo on Back Cover.

ABOUT THE AUTHOR

To have lived in the era of Charles Lindbergh, Amelia Earheart, Roscoe Turner, and the like was to insure a total fascination with flight for most boys like myself at that time. We read about flying, and dreamed about flying, as we saw the world's interest center around the adventures of transoceanic flyers, air racers, and flamboyant aviation personalities.

Our generation had an enthusiasm for the 'AIR' that never seemed to diminish.

The advent of World War II served to heighten this interest as hundreds of books and magazines told the stories of the exploits of famous fighter and bomber pilots.

These were exciting times, and for me to have a cousin in the United States Army Air Force who flew the B-26 Marauders on combat missions over Europe was every bit as exciting as the Air Battle of Britain or General Doolittle's bombing raid on Tokyo.

Now, nearly fifty years have passed since America's entry into World War II, and yet the memories of the courage and sacrifice of those who fought in that war have never dimmed. It was these memories which encouraged me to research and write the story of Captain John Quinn West, Jr.

It is a story which must be classed as historical fiction. Yet, it is as factual a story as this author could make it, for I was the kid called 'Jackie' in the early chapters who saw these things first hand many years ago.

POSTSCRIPT

The memories of childhood are strong indeed, and for me these days of youth were filled with the joy of summer vacations to the "Country" as we called it. Our family lived in a large city during the 1940's, and each summer I would always look forward to going to the home of my grandparents in Mississippi near Grenada, or to the home of my aunt and uncle in Sardis, Mississippi. Especially enjoyable was the fellowship with my cousins, who in those days were more like brothers and sisters than cousins. The ties of kinship were strong and the loss of a cousin would many times be a memory that would linger for generations.

For those of us who remember World War II it was the single great historic event of our lifetime. Our world revolved around it with family and loved ones serving in the armed forces overseas, and with our own military service. The many sacrifices we made on the home front for the war effort, all reminded us that we were in an age where events were being recorded that would live in the annuls of history forever. There was the joy of those coming home at the end of the war, and there were the vacant places in our lives left by the memory of those who failed to come back and were counted as heros.

A memorable feature of those days was the secrecy of military events, troop whereabouts, and the like. It was common for a wife or mother to know absolutely nothing of the whereabouts of their loved ones except for the possible postmarks on letters.

The men and women in military service were strictly encouraged to never mention military details in letters or telephone calls to home, and as strange as it may seem to us now, military censors were used to open servicemen's mail to check for any military information which would be quickly cut out, letters resealed and sent to their posted addresses. Moreover, it was common for those who lost loved ones in the war to receive a single line from the War Department saying, "We regret to inform you, your son was killed in action on August 1, 1944." No details of the way he died or where he died, no mention of his valor or gallantry in action. Sometimes his personal effects were mailed to parents and sometimes not. It was a time of humility and acceptance of the powers that be. It was a time when you bore your own grief and locked your sorrows within your heart.

For those servicemen who came home and were able to tell parents and wives about the many places they went and the varied experiences they had, it was a joyful revelation of what had happened during those long years when their son or daughter or spouse were in service. But, for those men who were killed in action those years would always be to their loved ones a puzzle, an enigma, and a mystery.

For most people there would never be a search of the facts, but for others there would be a nagging desire to find the truth of a kinsmans military service. This desire was the beginning of my quest for facts of the military life of Captain John Quinn West, Jr., a first cousin, an Air Force Pilot, and a gallant hero.

It was not an easy task, but it was a fulfilling one which covered a score of years and research in many states and in several countries.

It all began with a newspaper clipping in an old scrapbook which Captain West's mother had dutifully kept of all the events in her son's life. The school days, college days, and very little of his military service except a faded yellow news clipping of a plane with its crew which had belly landed in Normandy, France, on July 16, 1944. All of the crew members were named in the caption as well as their home towns. Twenty years had past since that plane crash, but I felt that some of these men certainly would be living in the same area and could be contacted.

I began to write the Chamber of Commerce in these towns to send me a xerox copy of names in directories which corresponded to the names of those for whom I was searching. I looked in libraries for telephone directories of those towns to find names which would be a family name of those I sought hoping that some relative might correspond with me and tell me where the man I looked for was living.

All to no avail. Hundreds of letters, and hundreds of replies from concerned people who wanted to help, but they had no information.

We had so little information to go on. Only his unit designation, the 397th bombardment group, and that he had served in England, and flew bombing missions over German occupied France.

It seemed to me that the 397th bomb group should be able to be found, but where to look for 397th members was a mystery to me. Should I run an ad in the paper, or should I advertise in an American Legion magazine, or would an ad in any Air Force Journal encourage a veteran of the 397th to correspond with me? It was frustrating to not know where to look for information.

I finally decided to place an ad in the Air Force Journal and was disappointed when no one answered. Months went by and a letter came from a man in England, Ken Fisher, who had seen a copy of the Journal and answered because he had some few facts about the 397th's base in England called Rivenhall. I was elated, also Ken had a friend who had researched the 397th for years and he suggested I write to him, a Mr. Bruce A Stait.

I immediately wrote to Mr. Stait and received a prompt reply. "Yes, he knew much about the 397th and had access to some few photos of the group which were lent to him by a 397th officer, Capt. James Snow, a photo officer of the group." The trail was getting definitely warm, and I could sense that Bruce was the key which would ultimately open doors to the research in which I was interested.

One of the most sharing and caring people with whom I have ever corresponded was Capt. James Snow. He answered my letter quickly, "Yes, he had photos of the group which I might photo copy and return to him. There were even several of Captain West and his crew beside his aircraft. Yes, he knew and remembered Captain West, a good pilot, and a sterling character. Yes, he knew how I could contact the 397th group members. He would send me a roster of the Reunion Group and I could write to West's crew members."

Wonder of wonders, how could I be so fortunate. All these years of searching was about to culminate in my contact with the very people who had flown with him in combat. I could hardly believe my good fortune. I began to write letters, piles of them to every one who might have had contact with him, but there were few answers. Several "By-Golly" crew members had died after the war, others were reluctant to correspond.

There was a reason for their reluctance which I fully understood and respected. They had been through a nightmare, their plane had been shot down by enemy fighters, and their Captain and Navigator had died in the crash. They had parachuted into enemy territory and were taken as P.O.W.'s. It was too painful for them to talk about it over the telephone and more painful to have to write about it.

I understood, but to whom would I turn now? There might be others who would know him. I had learned by bits and pieces that there were four squadrons in the 397th and West's squadron was the 598th. I could go down the roster and write to every man in the 598th squadron, surely someone would remember him, there must be some one who could say, "Yes, I knew him, he was a great pilot, I remember the time we flew together on a bomb mission over France." These were the things that I was hungry to hear.

It was years before all the pieces of the puzzle began to fit together. A scrap of information here, a photo of Capt'n West there. A book came in the mail one day. It was the 397th Bomb Group History, a book which was very valuable and rare. It had been printed a short time after the war and was considered a prize possession of the men who still had one. It had come from James Snow who was then lying in bed with terminal cancer, so weak that he was unable to even write but a few lines. He wanted me to have the book because of my interest in the 397th. I nearly wept to think that he would do this for me. It made all my efforts to research the group seem small in comparison to this man's effort to give me a treasure of the 397th even in his last moments.

There were letters from enlisted men in Captain West's squadron giving me some remembrance they had of him, a humorous story, or a serious happening. Joe Donzello, Bill Henry, Neil McGinnis, James Russell, and Ray Snow were all very helpful to me. Later West's crewmen Harold Zola, Chester Natanek, and Fred Daoust began to share many incidents of the "By-Golly" crew with me. The pieces were coming together and I was finally discovering where and how to research the 397th on every mission they flew.

The Albert Simpson Historical Research Library at Maxwell Air Force Base in Montgomery, Alabama, is the place where all Air Force historical records reside. If you go there you can find each day of the history of any unit and each mission that they flew. All in hard copy and in microfilm. It was a fantastic discovery for me. Now it would open up the missions, the crew lists, the loading lists, the squadron histories, the group histories, every tiny fact and statistic of their history was there. Even the bomb loads, the route to target, the results, the aircraft damage, the enemy fighter activity, the debriefing, and the reports from crew members about the mission were available for study. It was all too good to be true.

At about this time Bruce Stait was also corresponding with me and giving the results of his research. Mission lists with dates, places, lead pilots, serial numbers of the Marauders, their pilots, and names of their aircraft, all of which were vital to my research effort. I was continually astonished at the wealth of research which Bruce was willing to share with me.

Months previous to our initial correspondence I had published a small booklet about Captain West's Christian life and some few facts I knew about his military service. I mailed a copy to Bruce and was completely amazed as he related the following by return mail. When he and his brother were young boys their family lived near the Rivenhall Air Field called 'Chelmsford' by the airmen there at that time. The boys called it the Aerodrome, and many times they would ride their bicycles down the lane, go through the hedgerow onto the Field and talk to the crew of an airplane called "By-Golly."

They would watch this particular plane as it would take off and return from its missions, for the boys rather fancied it was their plane. One day the plane didn't come back from its mission and they always wondered what fate had befallen the ship and its crew. After forty years of not knowing, they had finally gotten their answer from a booklet about West's life.

Who would have ever believed the chance of this impossible occurrence that I should be able to contact the very man who had a personal knowledge of the aircraft "By-Golly?" Was it providential? Yes, it had to be. Let us reflect a moment.

"Forty years after World War II a man in America wants to learn more about an aircraft named "By-Golly." It was one plane among the many thousands of military aircraft which flew in England during the war. He places a tiny 'ad' in a magazine in America which a man in England accidentally sees. The man in England who saw the advertisement had a personal knowledge of this very aircraft, had conversed with the crew, and remembered the pilot, a captain with reddish blond hair and not a tall man. An exact description of Captain West."

A wealth of research over the years has produced a great knowledge of facts and statistics which have been invaluable to me in writing this story. But sadly, just the statistics do not provide the everyday small events in the life of the men, nor the many human interest stories, nor do they give insight into character. These must be discovered through years of personal interviews and good fortune in finding men who have lived through those trying times and can recall those events of forty years ago.

However, this was the exciting part for me to interview these airmen and let their stories take me back into the past to hear those long lost conversations and see a little of the lives of the men in the story, to possibly imagine flying with them on the missions and to share a part of their joys and sorrows, and to get just a small glimpse of the way it was back then, flying on the Wings of Courage.

EPILOGUE

The life of Captain John Quinn West, Jr. has been an inspiration to many people through the years, and although nearly fifty years have passed since his death, still his story continues to touch others with his life of dedicated Christian service. My purpose in writing his story has been that others might catch the glimmer of light which shown forth in his life and that they too might find the rich blessings and joy of a Christ centered life.

"Ye are the light of the world. A city that is set on a hill cannot be hid. Neither do men light a candle, and put it under a bushel, but on a candlestick; and it giveth light unto all that are in the house. Let your light so shine before men, that they may see your good works, and glorify your Father which is in heaven." Matthew 5:14-16

A gathering of officers with the Group Chaplain, Clarence Comfort. From left, Lt. Doss, Lt. Daoust, Capt. West, Lt. Cramer, and Capt. Comfort. In foreground is Lt. Crumm with "Jiggs."